THE HYPERPROOF MANIFESTO

ON LEGITIMACY, CONSENT & EXECUTABLE AUTHORITY

THE HYPERMODERN THEOREM
BOOK 2

DOM JOCUBEIT

REGISTERED OFFICES

Hypermodern Limited Company, 1209 Mountain Road PL NE, STE N, Albuquerque, NM 87110, USA

For details of our global editorial offices, customer services, and more information about Hypermodern products visit us at www.hypermodern.ai.

Hypermodern also publishes its books in a variety of electronic formats and by print-on-demand. Some content that appears in standard print versions of this book may not be available in other formats.

TRADEMARKS

Hypermodern and the Hypermodern logo are trademarks or registered trademarks of Hypermodern Limited Company and/ or its affiliates in the United States and other countries and may not be used without written permission. All other trademarks are the property of their respective owners. Hypermodern Limited Company is associated with some products mentioned in this book.

LIMIT OF LIABILITY/DISCLAIMER OF WARRANTY

For Chelsea

Who does not grant legitimacy without evidence.

———

Authority that cannot be enforced
is indistinguishable from permission.

CONTENTS

PART II

PART V

PART VI

THE HYPERMODERN THEOREM

This book addresses the second domain of the Hypermodern Theorem: authority.

If authoritative data establishes what exists, authority determines what may act upon it. Systems fail when this relationship is symbolic rather than enforceable, when permission, consent, or legitimacy are asserted without a mechanism that binds execution.

The Hyperproof Manifesto isolates this failure mode. It argues that institutions decay when authority is treated as a narrative property rather than an executable one. Proof without enforcement becomes ceremony. Consent without consequence becomes theater.

This book therefore advances a narrow claim: authority must be mechanically provable and directly coupled to execution, or it will eventually be bypassed, gamed, or ignored.

These claims are tested through Privacy First, a system designed to encode consent, proof, and legitimacy as enforceable primitives rather than social agreements. Privacy First is not presented as validation, but as a falsifiable implementation of the arguments made here.

This domain depends on the first.

Without unified, authoritative data, authority cannot be enforced without interpretation. Without enforceable authority, coordination cannot proceed without governance.

The Hyperproof Manifesto stands on its own. But within the Hypermodern Theorem, it is the binding layer.

If authority cannot be made executable without trust or appeal, the theorem collapses at this stage.

If it can, coordination becomes a solvable mechanical problem.

PREFACE

Manifestos are written when systems drift too far from their purpose.

They do not emerge from consensus. They emerge from friction, when existing language can no longer describe what is breaking, and incremental fixes only reinforce the problem.

This manifesto exists because digital systems have accumulated extraordinary power without developing an equivalent capacity for legitimacy.

Modern digital infrastructure can authenticate users, enforce policies, and automate outcomes at global scale. What it cannot do reliably is prove **why** a decision occurred, **who** committed to it, and **whether authority was exercised deliberately**.

Where proof is missing, inference takes over.

Inference invites surveillance. Surveillance erodes dignity. Dignity loss undermines trust.

This is not a failure of intent.

It is a failure of architecture.

POSITION

This text is not a call to optimize existing systems.

It does not propose better dashboards, smarter controls, or more comprehensive oversight. Those approaches assume the current model is correct and merely underpowered.

This manifesto rejects that assumption.

Its claim is simple: trust cannot scale where authority is implicit. Modern systems can act at extraordinary speed, but they cannot prove that power was exercised deliberately, lawfully, or with consent. In response, they compensate with surveillance, inference, and control.

This manifesto argues for a different foundation, one where decisions are explicit, consent is intentional, and legitimacy is cryptographically provable.

Legitimacy, consent, and executable authority are not features. They are prerequisites.

Anything less scales power without accountability.

Although this manifesto informs real systems and real platforms, it is not a product document. Platforms may implement these ideas. Standards may encode them. Regulations may reflect them.

The ideas themselves are prior.

CONTINUITY WITHIN THE HYPERMODERN THEOREM

This work follows *The Hyperstore Manifesto*.

Where that text examined the concentration of data and execution, this one examines the concentration of authority and legitimacy. Together, they describe a Hypermodern condition: systems that scale faster than the structures meant to govern them.

This manifesto addresses that imbalance by focusing on authority as a first-class, provable property of systems rather than an implied byproduct of control.

AUDIENCE

This manifesto is written for those who design systems that affect other people:

- Engineers and architects
- Security and identity practitioners
- Executives and operators
- Regulators and auditors
- Researchers shaping autonomous systems

It is also written for anyone who has felt the quiet discomfort of being governed by systems that never quite ask.

TONE

The claims in this book are deliberately strong. I make no apologies.

That is not because existing approaches are malicious or naive, but because legitimacy cannot be restored through incremental language. When authority is implicit, every optimization reinforces the same failure mode.

Manifestos do not ask for agreement.

They ask to be understood.

INVITATION

What follows is an argument for a different foundation, one where power is exercised only when it can be proven, and where trust is earned at the moment it is required.

You do not need to accept every conclusion to engage with the argument.

If this book sharpens your questions about proof, consent, or digital legitimacy, then it has done its work.

The rest of the book makes that case.

INTRODUCTION
THE END OF
IMPLICIT TRUST

For most of the history of computing, trust was implicit.

We trusted systems because they were closed. We trusted organizations because their boundaries were legible. We trusted data because it moved slowly, was expensive to copy, and lived inside institutional walls. Identity, authority, and accountability were bundled together by default.

That world no longer exists.

Artificial intelligence has not merely accelerated computation; it has dissolved the assumptions that made implicit trust workable. Decisions that once required deliberation now occur at machine speed. Authority is increasingly delegated to software. Actions compound faster than humans can supervise them, and the consequences of error, misuse, or abuse propagate instantly.

In this environment, trust can no longer be assumed, inferred, or reconstructed after the fact.

It must be **provable**.

This book examines what changes when we accept that premise as non-negotiable.

THE FAILURE OF ASSUMED LEGITIMACY

Modern digital systems rest on a quiet assumption: if an action occurred inside the system, it must have been legitimate.

Authentication verifies that someone logged in. Authorization checks that a policy permitted an action. Logging records that something happened.

Together, these mechanisms create the appearance of control. But they do not establish legitimacy. They cannot answer the questions that matter most when stakes are high:

- Who had the authority to make this decision?
- Was that authority valid at the moment the decision was made?
- Was the decision intentional, contextual, and bounded?

When these questions arise, organizations fall back on inference, testimony, and reconstruction. Logs are reviewed. Policies are cited. Processes are narrated after the outcome is known.

Trust becomes a story assembled after the fact.

This approach was fragile even before AI. It is untenable now.

AI CHANGES THE NATURE OF RISK

AI systems do not merely automate tasks; they amplify decisions.

They act continuously and adaptively, operate across organizational boundaries, and learn from historical data that may encode outdated assumptions or hidden biases. Increasingly, they initiate actions without direct human supervision.

In such systems, the cost of implicit trust is no longer linear. Small failures compound. Misconfigurations propagate. Authority drifts silently from humans to machines.

The central question is no longer whether a system is secure, compliant, or well-intentioned.

The question is whether its decisions are **legitimate**.

PRIVACY IS AN OUTCOME, NOT A FEATURE

In response to these risks, the industry has produced layers of tooling: stronger authentication, better monitoring, stricter compliance frameworks. These are necessary, but they are not sufficient.

Privacy, in particular, is often treated as a constraint to be managed or a requirement to be satisfied. Data is collected first and justified later. Controls are bolted on after systems are already operational.

This framing misses the point.

Privacy is not a feature. It is a consequence of how authority, identity, and decision-making are structured. Systems that cannot prove legitimacy inevitably over-collect, over-retain, and over-share. They compensate for uncertainty with data.

Systems that can prove legitimacy do not need to.

FROM CONTROL TO PROOF

What is required is not more control, but a different foundation.

One in which:

- Identity is cryptographic, not credential-based
- Authority is explicit, not inferred
- Decisions are first-class objects, not side effects
- Auditability is structural, not forensic

This shift is not incremental. It requires rethinking what systems are designed to prove, and when.

The core claim of this book is simple:

Trustworthy systems must be able to prove who decided what, under which authority, at the moment it mattered.

Everything that follows flows from that claim.

WHAT THIS BOOK IS — AND IS NOT

This is not a book about passwords, compliance checklists, or vendor comparisons. It is not a manifesto against AI, nor a technical manual for cryptography.

It is an argument for a different model of digital trust: one grounded in cryptographic proof, explicit authority, and verifiable decision-making.

A platform called *Privacy First* appears in this book. It is not the subject. It is a **falsifiable implementation**: a concrete system designed to test whether the ideas explored here can be made operational under real constraints.

HOW THIS BOOK PROCEEDS

We begin by examining the collapse of implicit trust in digital systems, how familiar mechanisms like authentication, authorization, and auditing came to substitute for proof, and why that substitution no longer works.

From there, we move toward a model where decisions are explicit, authority is signed, and governance is cryptographic by design.

If that future feels inevitable by the end of this book, then this introduction has done its job.

THE HYPERPROOF MANIFESTO

PART I

THE COLLAPSE OF
IMPLICIT TRUST

Modern digital systems appear trustworthy.

Users authenticate. Permissions are checked. Actions are logged. Dashboards report compliance status in reassuring shades of green. From the outside, everything suggests control, accountability, and order.

Yet when systems fail — when breaches occur, when fraud is discovered, when decisions are questioned — a familiar pattern emerges. No one can say, with certainty, who was actually responsible at the moment it mattered.

This is not because systems are poorly engineered. It is because they were never designed to prove trust in the first place.

Power scaled faster than legitimacy.

CHAPTER 1
THE ILLUSION OF DIGITAL TRUST

Most digital systems operate on an implicit contract:

If an action occurred inside the system, it must have been legitimate.

Authentication establishes that *someone* accessed the system. Authorization asserts that *some rule* allowed the action. Logging records that *something* happened.

These mechanisms do not prove trust. They assume it.

They assume that credentials were not stolen. That permissions were still appropriate. That context had not changed. That policies were correct. That intent was aligned with outcome.

As long as systems were slow, bounded, and human-operated, these assumptions held often enough to be useful. But they were never guarantees.

AUTHENTICATION IS NOT PROOF

Authentication answers a narrow question:

Can this actor present a valid credential?

It does not answer:

- whether the actor is present
- whether the actor understands the consequence of the action
- whether the actor intends the outcome
- whether the credential is being misused

Passwords, tokens, sessions, and even multi-factor authentication all suffer from the same limitation. They prove possession of a secret, not legitimacy of intent.

Even modern single sign-on systems and federated identity frameworks extend this model rather than replacing it. They centralize trust, but they do not make it provable.

AUTHORIZATION WITHOUT CONTEXT

Authorization systems answer a different question:

Is this action allowed under the current policy?

Policies, however, are static abstractions applied to dynamic reality. They encode assumptions about roles, responsibilities, and risk that inevitably drift over time.

A permission granted for one purpose quietly enables another. Temporary access becomes permanent. Emergency exceptions become routine. Authority accumulates invisibly.

When something goes wrong, the system can say the action was *permitted*. It cannot say it was *appropriate*.

LOGGING AS NARRATIVE

Logs are often treated as evidence.

In reality, they are stories.

A log records that an event occurred, not why it occurred, whether it was authorized intentionally, or whether the authority was valid at that moment. Logs capture sequence, not legitimacy.

During audits or investigations, logs are interpreted, correlated, and contextualized after the fact. Multiple plausible narratives are constructed. Responsibility is inferred, not proven.

This is why two competent teams can look at the same logs and reach different conclusions.

THE COMFORT OF CONTROL

The industry responds to these weaknesses by adding layers:

- stronger authentication
- more granular permissions
- richer logs
- real-time monitoring

Each layer increases the *appearance* of control. None address the underlying problem: the system still cannot prove that a decision was made by the right actor, under the right authority, with the right intent.

Control is substituted for proof.

WHEN TRUST FAILS

When trust collapses, organizations ask questions their systems were never designed to answer:

- Who approved this?
- Was that approval valid at the time?
- Did policy allow it, or did policy fail to prevent it?

The answers are reconstructed from memory, process documentation, and inference. Decisions are justified retroactively. Accountability becomes political.

At small scale, this is manageable. At machine scale, it is catastrophic.

THE LIMITS OF INCREMENTAL FIXES

Better passwords do not solve this problem.

Stronger MFA does not solve it.

More logging does not solve it.

These measures reduce risk at the margins, but they leave the core assumption intact: that trust can be inferred from system behavior.

It cannot.

THE MISSING QUESTION

Every system built on implicit trust avoids asking the question that matters most:

Who had the authority to decide this, at the moment the decision was made?

Until systems can answer that question with cryptographic certainty, trust will remain an illusion — persuasive, familiar, and fragile.

———

In the next chapter, we examine why identity alone cannot carry authority, and how authentication became a stand-in for proof.

CHAPTER 2
IDENTITY
WITHOUT PROOF

Identity sits at the center of modern digital trust.

Every access request, policy evaluation, and audit trail begins with the same assumption: if the system knows *who* you are, it can decide *what* you are allowed to do.

This assumption is so deeply embedded that it is rarely questioned. Identity systems are treated as neutral infrastructure, a prerequisite rather than a design choice.

But identity, as it is implemented today, cannot carry the weight placed upon it.

HOW IDENTITY BECAME A PROXY FOR AUTHORITY

In early systems, identity was local and limited. A user account mapped to a small set of actions inside a bounded environment. Authority was implicit and scope was narrow.

As systems grew, identity expanded to fill gaps it was never designed to address:

- Single sign-on promised convenience across systems
- Federation promised portability across organizations
- Centralized directories promised consistency

Over time, identity became the *anchor* for authority. If a user could be reliably identified, the system assumed their actions could be trusted.

This substitution worked only as long as identity remained tightly coupled to context, intent, and supervision. At scale, that coupling breaks.

CREDENTIALS ARE NOT PRESENCE

Identity systems authenticate credentials, not people.

A password, token, or session proves that a secret was presented. It does not prove:

- that the legitimate holder is present
- that the action is intentional
- that the context is understood

Even when identity is strongly verified at enrollment, the act of authentication is disconnected from the act of decision-making. Sessions persist. Tokens travel. Authority lingers long after presence ends.

This gap is not accidental. It is structural.

FEDERATION MULTIPLIES ASSUMPTIONS

Federated identity systems extend trust across boundaries by design.

They allow one system to assert identity on behalf of another. This reduces friction, but it also multiplies assumptions:

- that the upstream system verified the user correctly
- that its policies align with downstream risk
- that its context is still valid

When these assumptions fail, failures propagate silently. Downstream systems inherit authority without visibility into how it was established.

Identity becomes a chain of trust that no participant fully controls.

THE OVERLOADING OF IDENTITY

Modern identity systems are asked to do too much.

They are expected to:

- authenticate users
- convey roles and attributes
- enforce policy
- satisfy compliance
- explain decisions after the fact

Each responsibility stretches identity further from its original purpose. The result is an object that represents *who someone is, what they might be allowed to do,* and *why an action occurred;* without being able to prove any of these conclusively.

Identity becomes a container for assumptions.

PROFILES ARE NOT PROOF

To compensate for uncertainty, systems enrich identity with data:

- attributes
- claims

- behavioral signals
- historical activity

The profile grows. Context is inferred. Risk is scored.

Yet none of this establishes legitimacy. It increases confidence without increasing certainty. When decisions are challenged, profiles explain behavior; they do not prove authorization.

This is why identity-centric systems tend to over-collect data. Uncertainty is answered with information.

IDENTITY AND THE ILLUSION OF CONTROL

From the perspective of operators, identity-centric systems feel powerful.

Dashboards display users, roles, and access graphs. Changes propagate instantly. Revocation appears decisive.

But this sense of control is retrospective. Identity systems are excellent at describing the present state. They are poor at proving past authority.

When asked whether an action was legitimate *at the moment it occurred*, identity systems can only answer probabilistically.

THE WRONG QUESTION

Identity answers the question:

Who is this?

Trust requires an answer to a different question:

Who had the authority to decide this?

These questions are related, but they are not interchangeable. Authority is contextual, bounded, and temporal. Identity is not.

Treating identity as a substitute for authority conflates recognition with legitimacy.

THE CONSEQUENCE

As long as identity remains the primary trust primitive:

- authority will drift
- decisions will be implicit
- audits will reconstruct narratives
- privacy will erode

The problem is not weak identity. It is misplaced responsibility.

————

In the next chapter, we examine why authorization systems inherit these flaws, and how access control became a brittle stand-in for governance.

CHAPTER 3
AUDITABILITY
AS THEATER

When systems cannot prove trust, they compensate by explaining it.

Audit logs, compliance reports, and review processes exist to answer questions that the system itself cannot. They reconstruct events, correlate signals, and assemble narratives intended to demonstrate that actions were legitimate.

This practice is so normalized that it is rarely questioned. Audits are treated as evidence of control. Logs are treated as facts. Compliance becomes a proxy for correctness.

But auditability, as it is commonly implemented, is not proof. It is performance.

LOGS RECORD EVENTS, NOT AUTHORITY

A log entry tells you that something happened:

- an account was accessed
- a permission was evaluated
- a transaction was executed

It does not tell you:

- whether the actor had valid authority at that moment
- whether the decision was intentional
- whether the context justified the action

Logs capture sequence. Authority exists outside the log.

During an investigation, logs are read backward from an outcome. Events are stitched together into a plausible story. The same data can support multiple interpretations, depending on what the reviewer already believes.

This is not a tooling limitation. It is an information gap.

AUTHORIZATION EXPLAINS OUTCOMES, NOT DECISIONS

Authorization systems are designed to answer a present-tense question:

Is this action allowed right now?

When asked later whether an action *should have been allowed,* they can only replay policy logic against historical data. This replay assumes that:

- the policy was correct
- the inputs were complete
- the evaluation context was preserved

In practice, none of these assumptions hold reliably. Policies change. Context evaporates. Inputs are normalized or discarded.

Authorization can explain why a system permitted an action. It cannot prove that permitting it was legitimate.

COMPLIANCE AS RECONSTRUCTION

Compliance frameworks rely on evidence produced after the fact:

- access reviews
- change logs
- approval records
- attestations

These artifacts are assembled to demonstrate that appropriate controls *likely* existed. They do not establish that a specific decision was authorized by the right actor, under the right authority, at the right time.

This is why compliance exercises often feel adversarial. Auditors ask questions systems were never designed to answer. Organizations respond with documentation, screenshots, and explanations.

Compliance becomes a negotiation.

THE NARRATIVE TRAP

Once trust is reconstructed narratively, accountability becomes subjective.

Different stakeholders emphasize different signals. Legal teams look for policy alignment. Security teams look for anomalies. Operators look for procedural adherence.

Without a cryptographic anchor, none of these perspectives can claim finality. Responsibility is inferred. Intent is assumed. Legitimacy is debated.

In high-stakes environments, this ambiguity is not merely inconvenient. It is dangerous.

WHY MORE LOGGING DOES NOT HELP

The instinctive response to audit failure is to log more.

Richer telemetry, longer retention, finer-grained events, all increase observability. None address the underlying absence of proof.

More data sharpens narratives. It does not eliminate them.

In fact, excessive logging often worsens the problem. The volume of data increases the surface for interpretation, while the signal of authority remains missing.

AUDITABILITY VS. VERIFIABILITY

Auditability answers the question:

Can we explain what happened?

Verifiability answers a different question:

Can we prove that this decision was legitimate?

Most systems optimize for the former and quietly hope it implies the latter.

It does not.

WHEN AUDITS FAIL

When an organization cannot prove legitimacy, audits collapse into trust exercises:

- Who do we believe?
- Which explanation is more credible?
- What level of risk is acceptable?

These are governance questions, but they are answered with process and persuasion rather than evidence.

At scale, this becomes unsustainable.

THE MISSING ARTIFACT

What audits lack is not visibility, context, or data.

They lack a durable artifact that captures:

- a decision
- the authority under which it was made
- the actor who made it
- the moment it occurred

Without this artifact, auditability will remain theatrical, convincing until it is challenged, and fragile when it matters most.

———

In the next chapter, we examine why authentication and authorization cannot supply this artifact, and why authority must be modeled explicitly.

CHAPTER 4
AUTHENTICATION IS NOT AUTHORIZATION

Authentication and authorization are often discussed together, implemented together, and audited together.

They are not the same thing.

Authentication answers the question of identity. Authorization answers the question of permission. Conflating the two has allowed systems to scale, but it has also obscured a critical gap: neither establishes authority.

THE CONVENIENCE OF COUPLING

In most systems, authentication precedes authorization so reliably that the distinction feels academic. A user logs in, a policy is evaluated, and an action is either permitted or denied.

This coupling is convenient. It allows identity to stand in for authority and policy to stand in for intent.

Over time, this convenience hardened into assumption.

WHAT AUTHENTICATION ACTUALLY PROVES

Authentication proves that a credential was presented successfully.

Even in its strongest modern forms, hardware-backed keys, biometrics, multi-factor verification; authentication establishes only that:

- a key is valid
- a challenge was answered
- an account is recognized

It does not establish why an action is being taken, whether it is appropriate, or whether authority should be exercised at that moment.

Authentication proves access, not legitimacy.

WHAT AUTHORIZATION ACTUALLY EVALUATES

Authorization systems evaluate rules.

They take inputs, identity attributes, roles, context signals, and apply policy logic to decide whether an action is allowed.

This evaluation is deterministic and repeatable. It is also indifferent to intent.

If the inputs satisfy the policy, the action is permitted. Whether the actor understands the consequence, intends the outcome, or should exercise authority in that context is outside the model.

Authorization enforces consistency, not judgment.

THE STATIC NATURE OF ROLES

Role-based access control remains the dominant authorization model because it is simple and legible.

Roles encode assumptions about responsibility. They are granted, inherited, and revoked according to process.

But roles are static representations applied to dynamic reality. They age poorly. As systems evolve, roles accumulate permissions long after their original rationale has faded.

A role answers the question:

What might this actor be allowed to do?

It does not answer:

Should this actor exercise authority now?

ATTRIBUTE EXPLOSION

To address the rigidity of roles, systems layer attributes and conditions:

- time of day
- location
- device posture
- risk scores

This produces more nuanced authorization decisions, but it does not resolve the underlying issue. Attributes refine permission; they do not establish legitimacy.

As policies grow more complex, they become harder to reason about and harder to audit. Authority remains implicit, buried inside evaluation logic.

AUTHORIZATION WITHOUT ACCOUNTABILITY

When authorization permits an action, responsibility diffuses.

Was the policy wrong? Were the inputs incomplete? Was the context misunderstood? Each explanation is plausible, and none are provable.

Authorization systems can explain *how* a decision was reached. They cannot explain *why authority was exercised*.

This distinction matters when consequences are real.

THE ABSENCE OF INTENT

Neither authentication nor authorization captures intent.

Intent is not a role, an attribute, or a policy outcome. It is a deliberate choice made by an actor who understands the implications of that choice.

Without intent, systems confuse capability with legitimacy.

THE MISSING LAYER

Between authentication and authorization lies an unmodeled layer:

- the explicit act of deciding
- the acknowledgment of consequence
- the exercise of authority

Because this layer is missing, systems infer it from outcomes. They assume that permitted actions were intended and legitimate.

This assumption is increasingly unsafe.

WHY THIS MATTERS NOW

As AI systems act faster, broader, and with less supervision, authorization decisions increasingly trigger irreversible effects.

In such environments, permission is not enough. Authority must be explicit, provable, and bounded.

Authentication and authorization can support this model. They cannot replace it.

TOWARD EXPLICIT AUTHORITY

To move beyond implicit trust, systems must distinguish:

- identity from authority
- permission from decision
- evaluation from intent

This requires modeling decisions explicitly, as artifacts that can be signed, recorded, and verified.

———

In the next chapter, we examine why decisions themselves must become first-class objects, and what changes when systems are built around them.

PART II

DECISIONS AS THE SOURCE OF AUTHORITY

All systems exist to make decisions.

They decide who can access what, which transactions proceed, how resources are allocated, and when actions are triggered. Every state change, every side effect, every outcome is the result of a decision, whether explicit or implied.

And yet, modern systems almost never model decisions directly.

Systems learned to act before they learned to decide.

CHAPTER 5
DECISIONS ARE THE MISSING PRIMITIVE

In most architectures, decisions are embedded inside other constructs:

- conditional logic in code
- policy evaluations in authorization engines
- workflow transitions
- automated rules and heuristics

The decision itself is never represented as an object. It is inferred from the outcome.

If a record was updated, a decision must have been made.

If access was granted, someone must have approved it.

If funds were transferred, authority must have existed.

These inferences are convenient. They are also fragile.

OUTCOMES ARE NOT DECISIONS

An outcome tells you *what happened*.

A decision tells you:

- who chose it
- under what authority
- at what moment
- with what understanding of consequence

When systems conflate outcomes with decisions, they lose the ability to reason about legitimacy. They can replay events, but they cannot verify intent.

This distinction becomes critical when decisions are disputed.

WHY DECISIONS WERE IGNORED

Decisions were historically ignored as first-class objects for practical reasons.

Early systems were constrained by storage, computation, and network latency. Modeling every decision explicitly would have been expensive and unnecessary. Human oversight filled the gaps.

As systems scaled, these shortcuts became architectural habits. Decisions remained implicit even as consequences grew more severe.

The result is a mismatch between system power and system accountability.

THE COST OF IMPLICIT DECISIONS

When decisions are implicit:

- authority drifts over time
- intent is assumed
- accountability diffuses
- audits reconstruct narratives

Systems behave as if legitimacy were a property of outcomes rather than choices.

This inversion is subtle, but profound.

DECISIONS AS FIRST-CLASS OBJECTS

Treating decisions as first-class objects changes the architecture fundamentally.

A first-class decision:

- is explicitly defined
- is deliberately made
- is cryptographically bound to an actor
- is recorded immutably

State changes become *derivatives* of decisions, not evidence of them.

FROM EVALUATION TO COMMITMENT

Most authorization systems perform evaluation.

They answer whether an action satisfies policy. They do not require an actor to commit to the consequences of that action.

A decision requires commitment.

It is an explicit acknowledgment that authority is being exercised and that the outcome is intended.

This distinction is what separates automation from governance.

INTENT AS A TECHNICAL CONCEPT

Intent is often treated as psychological or legal.

In practice, intent can be modeled technically:

- a unique challenge
- a verified presence
- a deliberate signature

When intent is captured at the moment of decision, legitimacy becomes verifiable rather than inferred.

DECISIONS ENABLE GOVERNANCE

Once decisions are explicit:

- authority can be scoped and bounded
- responsibility can be attributed precisely
- audits become deterministic
- disputes resolve to evidence

Governance stops being procedural and becomes structural.

THE PRIMITIVE EVERYTHING ELSE DEPENDS ON

Identity answers *who*.

Permissions answer *what might be allowed*.

Policies answer *under which rules*.

Decisions answer *what was actually authorized*.

Without explicit decisions, trust systems remain incomplete.

WHAT COMES NEXT

Recognizing decisions as the missing primitive raises a new question:

How are decisions made explicit, provable, and immutable?

The answer requires a new construct, one that captures intent, authority, and consequence in a single, verifiable artifact.

———

In the next chapter, we introduce intent, presence, and cryptographic proof; the foundations required to make decisions real.

CHAPTER 6
INTENT, PRESENCE, AND CRYPTOGRAPHIC PROOF

If decisions are the missing primitive, intent is the missing signal.

Without intent, systems cannot distinguish between an action that was merely possible and one that was deliberately authorized. They can evaluate rules, but they cannot establish commitment. They can observe outcomes, but they cannot prove legitimacy.

To make decisions real, systems must be able to capture intent at the moment it is exercised.

WHY INTENT CANNOT BE INFERRED

Intent is often treated as something that can be reconstructed from behavior.

If an action occurred, intent is assumed. If it was repeated, confidence increases. If it aligns with role and policy, legitimacy is inferred.

This approach fails precisely when stakes are high. Automated systems, compromised credentials, delegated authority, and machine-initiated actions all produce outcomes without human intent.

Inference collapses under ambiguity.

PRESENCE AS A PREREQUISITE

Intent requires presence.

An actor cannot meaningfully intend an outcome unless they are present at the moment of decision, aware of context, and capable of consenting to consequence.

Most authentication systems do not establish presence. Sessions persist beyond attention. Tokens operate without supervision. Authority outlives awareness.

Presence must be proven, not assumed.

THE LIMITS OF BEHAVIORAL SIGNALS

Modern systems attempt to approximate intent using signals:

- click patterns
- timing analysis
- risk scores
- anomaly detection

These signals can reduce fraud. They cannot prove consent or authorization. At best, they indicate likelihood. At worst, they create false confidence.

Intent cannot be probabilistic.

CRYPTOGRAPHIC COMMITMENT

Cryptography offers a property that behavioral inference cannot: commitment.

A cryptographic signature is an explicit act. It binds an actor to a specific statement at a specific moment. It cannot be forged retroactively, replayed invisibly, or altered without detection.

When an actor signs a decision, they are not merely authenticated, they are committing authority.

This is the difference between recognition and proof.

PRESENCE THROUGH CHALLENGE AND RESPONSE

To bind intent to presence, systems must ensure that:

- the actor is available at the moment of decision
- the decision cannot be precomputed or replayed
- the actor acknowledges the context of the choice

Challenge–response protocols achieve this by design. A unique challenge is issued. A response is generated in real time. The response is valid only for that moment.

Presence becomes a cryptographic fact.

WHY HARDWARE MATTERS

Software-only keys are vulnerable to delegation, automation, and exfiltration.

Hardware-backed keys introduce a physical boundary. They require user interaction. They resist silent use. They make intent difficult to fake at scale.

This is not about biometrics or convenience. It is about ensuring that authority cannot be exercised without friction and awareness.

SIGNATURES AS EVIDENCE

A signed decision is durable.

It can be verified independently. It can be audited without trust in the system that produced it. It can be evaluated years later with the same certainty as at the moment it was created.

This durability is what logs and policies lack.

FROM IDENTITY TO INTENT

When systems rely on identity, they ask:

Who is this?

When systems rely on cryptographic intent, they ask:

Who chose this?

This shift is subtle, but transformative. Authority moves from accounts to actions. Trust moves from systems to proofs.

MAKING DECISIONS REAL

With presence established and intent signed, decisions become real artifacts:

- explicit
- verifiable
- immutable

They can be recorded, referenced, and reasoned about directly.

This is the final prerequisite for modeling authority explicitly.

———

In the next chapter, we introduce the mandate: a signed, bounded, and immutable decision that serves as the foundation of cryptographic governance.

CHAPTER 7
WHAT A
MANDATE IS

If intent can be proven, authority can be made explicit.

The previous chapter established three requirements for legitimate decisions:

- the actor must be present
- intent must be deliberate
- commitment must be cryptographically provable

What remains is to bind these requirements into a durable construct that systems can reason about directly.

That construct is the mandate.

THE PROBLEM THIS CHAPTER SOLVES

Most systems cannot answer a simple question with certainty:

> Who had the authority to make this decision, at
> the moment it was made?

They approximate. They infer. They reconstruct. But they cannot *prove*.

This is not a logging failure or an audit tooling problem. It is an architectural omission. Modern systems do not model **decisions** as first-class objects. They model state transitions, API calls, and permissions, and then attempt to infer intent after the fact.

A **mandate** exists to correct that omission.

DEFINITION: MANDATE

A **mandate** is a cryptographically signed declaration of intent that authorizes a specific decision under explicit constraints, at a specific point in time, by a verifiable actor.

Formally, a mandate has the following properties:

1. **Intentional** — it represents a deliberate choice, not an implicit side effect
2. **Authenticated** — it is signed using a strong, non-exportable cryptographic key
3. **Bounded** — it applies to a defined scope, context, and set of outcomes
4. **Temporal** — it is anchored to a moment and cannot be replayed
5. **Immutable** — once recorded, it cannot be altered without detection

A mandate is not:

- a permission
- a role
- a policy
- a workflow step

Those are *derivatives*. A mandate is the root.

DECISIONS AS FIRST-CLASS OBJECTS

In conventional systems, decisions are implicit. They occur when:

- an API endpoint is invoked
- a database row is updated
- a workflow transitions state

The decision is *assumed* to have occurred because the system changed.

In a mandates-based system, this assumption is reversed.

State changes occur *because* a decision was explicitly made, signed, and recorded.

This inversion has three consequences:

1. **Authority becomes provable**
2. **Auditability becomes deterministic**
3. **Governance becomes composable**

THE MANDATE LIFECYCLE

Every mandate follows a strict lifecycle:

1. **Proposal** — a decision is defined, including its choice architecture
2. **Challenge** — a unique cryptographic challenge is issued
3. **Verification** — the actor proves presence and intent (e.g., passkey)
4. **Signature** — the decision is cryptographically signed

5. **Recording** — the signed mandate is immutably stored
6. **Derivation** — system state is derived from the mandate

Notably absent from this list:

- sessions
- shared secrets
- implicit trust

WHY MANDATES ARE NOT PERMISSIONS

Permissions describe *what might be allowed*.

Mandates prove *what was explicitly authorized*.

A permission model answers:

> Could this actor have done this?

A mandate model answers:

> Did this actor authorize this, under these conditions, at this time?

This distinction is the difference between **capability** and **legitimacy**.

BINARY DECISIONS ARE A DEGENERATE CASE

Traditional approval workflows reduce decision-making to a binary:

- Approve
- Reject

This is not wrong, but it is unnecessarily restrictive.

A mandate supports:

- binary approvals
- multi-choice selections
- ranked preferences
- conditional routing
- quorum-based outcomes

Binary approval is simply the smallest possible choice set.

MANDATES VS. WORKFLOWS

Workflows describe *process*.

Mandates describe *authority*.

A workflow can exist without legitimacy.

A mandate cannot.

This separation allows:

- workflows to change without invalidating authority
- authority to be verified independent of execution

FAILURE MODES WITHOUT MANDATES

Systems without explicit mandates fail in predictable ways:

- **Silent escalation** — authority accumulates invisibly
- **Retroactive justification** — logs are interpreted to fit outcomes
- **Audit ambiguity** — multiple actors appear responsible
- **Policy drift** — permissions outlive their rationale

Mandates eliminate these by construction.

MANDATES AS GOVERNANCE PRIMITIVES

Once decisions are explicit and signed, governance becomes programmable:

- approvals become attestations
- votes become cryptographic facts
- role changes become derived outcomes

This is not process automation.

It is **legitimacy automation**.

WHAT THIS ENABLES

With mandates in place:

- Authority is provable without trust
- Audits are queries, not investigations
- Disputes resolve to evidence, not narratives
- Systems can explain themselves

This is the minimum viable foundation for trust in the AI era.

―――――

In the next chapter, we examine how choice architecture expands beyond binary approval, and why unlimited decision spaces are necessary for real governance.

CHAPTER 8
CHOICE ARCHITECTURE WITHOUT LIMITS

Once decisions are explicit, a new constraint becomes visible.

Most systems only know how to decide *yes* or *no*.

Approval or rejection. Allow or deny. Pass or fail.

This binary framing has shaped the design of workflows, governance processes, and security controls for decades. It is simple, automatable, and easy to reason about. It is also profoundly limiting.

THE TYRANNY OF BINARY DECISIONS

Binary decisions are an artifact of implementation, not a property of reality.

Human organizations rarely decide in absolutes. They choose between options, rank priorities, negotiate trade-offs, and route decisions based on context. They vote, defer, escalate, and revisit.

When systems force these processes into yes/no approvals, complexity does not disappear. It leaks.

WHERE COMPLEXITY GOES TO HIDE

When a decision has more than two legitimate outcomes,
binary systems compensate by layering:

- parallel approval chains
- conditional workflows
- out-of-band discussions
- informal consensus

The system records the final outcome, but the real decision happens elsewhere. Authority fragments across
email threads, meetings, and undocumented conver-
sations.

What remains in the system is a shadow of the decision, not
the decision itself.

APPROVAL IS NOT CHOICE

An approval answers a narrow question:

> May this proceed?

A choice answers a broader one:

> Which of these outcomes should occur?

Conflating the two forces organizations to express preference
as permission. The result is brittle workflows that encode
governance logic indirectly and obscure intent.

DECISIONS WITH MORE THAN TWO OUTCOMES

Once decisions are modeled explicitly, there is no technical
reason to restrict them to two options.

A decision can legitimately involve:

- multiple candidates
- ranked preferences
- weighted votes
- quorum requirements
- conditional branches

The decision artifact remains the same. Only the choice set changes.

CHOICE ARCHITECTURE AS GOVERNANCE

The structure of choices shapes behavior.

Which options are available, how they are presented, and how outcomes are resolved all influence legitimacy. This is not a user-interface concern; it is a governance primitive.

By making choice architecture explicit, systems expose governance assumptions that were previously implicit.

VOTING WITHOUT FRAGILITY

Traditional voting systems rely on procedural trust: who ran the vote, who counted the ballots, who controlled the system.

When votes are explicit decisions, each choice can be individually signed, verified, and audited. The aggregate outcome becomes reproducible rather than asserted.

Trust shifts from process to proof.

QUORUMS, THRESHOLDS, AND AUTHORITY

Many decisions are not about majority preference, but about sufficiency of authority.

- a minimum number of approvers
- representation from specific roles
- time-bounded participation

These constraints belong to the decision itself, not to external workflow logic. Encoding them directly into the decision model makes authority explicit and reviewable.

ROUTING AS A DECISION OUTCOME

Some decisions do not select an end state; they select a path.

Escalation, delegation, and deferral are outcomes in their own right. Treating them as first-class choices allows systems to evolve dynamically without losing auditability.

ELIMINATING THE SIDE CHANNELS

When systems can represent real choices, informal side channels lose their necessity.

Discussions still happen, but the authoritative moment, the binding choice, occurs inside the system, captured as intent and commitment.

Governance becomes observable without becoming brittle.

FROM WORKFLOWS TO DECISIONS

Workflows describe *how* something happens.

Decisions describe *what was chosen*.

When decisions are explicit, workflows become execution details rather than sources of authority. Systems regain the ability to reason about legitimacy independent of process.

WHY THIS MATTERS FOR AI-DRIVEN SYSTEMS

As AI systems increasingly propose actions rather than execute them, the number of possible outcomes expands.

Binary approvals collapse nuance. Rich choice architectures preserve human authority while allowing automation to scale.

Decisions become the interface between human intent and machine capability.

THE FOUNDATION FOR MANDATES

Unlimited choice is not chaos.

When choices are bounded, signed, and verified, flexibility increases without sacrificing control. This is the foundation on which mandates operate: explicit authority exercised over explicit options.

———

In the next chapter, we examine how mandates bind decisions to scope, time, and consequence, and why authority must always be limited to remain legitimate.

CHAPTER 9
BOUNDED AUTHORITY
SCOPE, TIME, AND CONSEQUENCE

Authority that is not bounded is not authority.

It is exposure.

Every exercise of power carries risk. The legitimacy of that power depends not only on who exercises it, but on how far it extends, how long it lasts, and what consequences it can produce.

Modern systems are exceptionally good at granting authority. They are far less capable of limiting it.

THE PROBLEM OF STANDING AUTHORITY

Most systems grant authority in advance.

Roles are assigned. Permissions are accumulated. Access persists until explicitly revoked. This model assumes that authority, once granted, remains appropriate indefinitely.

Reality is less stable.

Responsibilities change. Context shifts. Threat models evolve. Yet standing authority quietly remains.

What was once justified becomes invisible.

AUTHORITY DRIFT

Over time, permissions accrete.

Temporary needs become permanent grants. Exceptional access becomes routine. Emergency privileges are never unwound.

This drift is rarely malicious. It is structural. Systems lack a native concept of expiry or purpose.

Authority expands because nothing constrains it.

SCOPE AS A FIRST-CLASS CONSTRAINT

Authority should be specific.

Not "may act," but:

- may act on *this resource*
- may perform *this operation*
- may affect *this domain*

When scope is implicit, authority is ambiguous. When scope is explicit, authority becomes reviewable.

A decision without scope is an open-ended permission.

TIME AS A BOUNDARY

Authority should decay.

Time-bound authority reflects a simple truth: legitimacy expires.

Yet most systems treat time as an external concern, a scheduled review, a manual cleanup, a governance checklist item.

When time is embedded directly into the decision, authority ends by design rather than by policy.

CONSEQUENCE AWARENESS

Authority is meaningful only in relation to consequence.

Approving a document is not the same as transferring funds. Granting read access is not the same as deleting records.

Systems that fail to model consequence flatten risk. Actors exercise authority without clear visibility into impact.

Legitimate decisions require consequence awareness at the moment of commitment.

BOUNDED AUTHORITY AS LEGITIMACY

When scope, time, and consequence are explicit:

- authority is exercised deliberately
- excess privilege is prevented structurally
- reviews become concrete rather than procedural

Boundaries are not restrictions on trust. They are the conditions that make trust rational.

REVOCATION IS NOT A BOUNDARY

Revocation is reactive.

It assumes authority has already been exercised, or could be at any moment. It relies on detection, response, and cleanup.

Bounded authority is proactive. It limits what can happen even if credentials are misused or systems fail.

This distinction matters under adversarial conditions.

MANDATES AS BOUNDED AUTHORITY

A mandate is not a blanket grant.

It is a decision with:

- defined scope
- explicit duration
- acknowledged consequence

Because mandates are explicit, these boundaries are not inferred. They are enforced by construction.

PREVENTING SILENT ESCALATION

When authority is bounded, escalation requires a new decision.

No additional scope, time, or consequence can be assumed. Each expansion must be deliberate, present, and signed.

This prevents silent escalation and makes power visible.

WHY THIS CHANGES EVERYTHING

Bounded authority transforms governance from a best practice into an invariant.

Systems no longer ask whether authority *should* be limited. They require that it is.

Legitimacy becomes a property of design, not discipline.

_____

In the next chapter, we examine how mandates integrate with roles, access control, and grants, and why roles should derive from decisions, not the other way around.

CHAPTER 10
ROLES ARE DERIVED, NOT GRANTED

Roles are one of the most enduring abstractions in access control.

They are familiar, convenient, and widely understood. Titles map cleanly to responsibilities. Permissions cluster naturally. Systems scale by assigning roles rather than reasoning about every action individually.

And yet, roles are not authority.

THE SHORTCUT THAT BECAME A FOUNDATION

Roles emerged as a shortcut.

Instead of deciding who could do what every time, systems grouped permissions under a name and reused it. This dramatically simplified administration and enabled early systems to scale.

Over time, the shortcut hardened into architecture. Roles became the primary mechanism by which authority was expressed.

The original decision, *why* this role existed and *who* should hold it, faded into the background.

ROLES DESCRIBE CAPABILITY, NOT LEGITIMACY

A role answers a limited question:

> What actions might this actor be able to perform?

It does not answer:

> Under what authority should those actions be taken?

Roles describe potential. They do not encode intent, scope, time, or consequence. As a result, they can never establish legitimacy on their own.

STANDING PERMISSION REVISITED

Because roles are persistent, they reintroduce the problem of standing authority.

Once assigned, a role quietly authorizes actions long after the context that justified it has changed. Reviews become procedural. Exceptions become permanent.

The system cannot distinguish between authority that is active and authority that is merely present.

WHY REMOVING ROLES FAILS

Some systems attempt to solve this by eliminating roles entirely.

This rarely works. Roles serve real cognitive and organizational functions. Humans reason in terms of responsibility and function, not raw permissions.

The failure is not the existence of roles. It is their position in the hierarchy.

AUTHORITY MUST PRECEDE ROLE

Authority should flow from decisions, not from titles.

A role should exist because a mandate requires it. An actor should hold a role because a decision explicitly grants it, with defined scope and duration.

In this model, roles are *derived artifacts*.

They materialize authority rather than creating it.

GRANTS AS ROLE MATERIALIZATION

A grant is the concrete expression of a mandate.

It translates a bounded decision into operational reality:

- assigning a role
- enabling a permission
- activating access

Because the grant is derived, it inherits the mandate's constraints. When the mandate expires or is revoked, the grant dissolves automatically.

TEMPORARY BY DEFAULT

In a mandate-driven system, permanence is the exception.

Roles are granted for a purpose, within a scope, and for a

limited time. Renewal requires a new decision, not a background job.

This aligns system behavior with how authority actually works in the real world.

REVIEWS BECOME EVIDENCE-BASED

When roles derive from mandates, reviews change character.

Instead of asking whether a role assignment still "makes sense," reviewers examine the decision that justified it. Scope, time, and consequence are explicit.

Governance becomes concrete.

AUTOMATION WITHOUT DRIFT

Derived roles enable automation without authority drift.

Systems can provision access just-in-time, revoke it automatically, and adapt to context; all *without accumulating silent privilege.*

Automation executes decisions. It does not create them.

REFRAMING RBAC

This model does not reject role-based access control. It reframes it.

RBAC becomes an execution layer, not a trust layer. Roles are implementation details, not sources of legitimacy.

This preserves the strengths of RBAC while eliminating its most dangerous failure mode.

AUTHORITY HAS A SOURCE

When roles are derived, authority becomes traceable.

Every permission points back to a decision. Every role assignment has provenance. Every access can be explained without narrative reconstruction.

This is the difference between administration and governance.

───────

In the next chapter, we examine how mandates, grants, and decisions form a unified decision fabric, and why security architecture must be built around it rather than layered on top.

PART III

THE DECISION FABRIC

Once decisions are explicit, bounded, and provable, they stop being isolated events.

They begin to form a fabric.

A decision fabric is the connective structure that links intent, authority, execution, and outcome into a coherent whole. It is not a workflow engine, a policy layer, or an audit log. It is the substrate on which governance is built.

Authority is real only when it can be proven.

CHAPTER 11
THE DECISION FABRIC

In a decision-centric system, every meaningful exercise of authority is represented as a node.

Approvals, votes, grants, delegations, escalations — all are instances of the same underlying construct: a decision made by an actor, at a moment in time, with defined scope and consequence.

The type of decision varies. The structure does not.

RELATIONSHIPS CREATE MEANING

A single decision rarely exists in isolation.

Decisions reference other decisions:

- a grant derives from a mandate
- an approval enables a transaction
- an escalation supersedes a prior choice

These relationships form edges between nodes. Together, nodes and edges create a graph of authority.

Meaning emerges from structure.

PROVENANCE WITHOUT NARRATIVES

Traditional systems reconstruct provenance by stitching together logs, tickets, and policy evaluations.

In a decision fabric, provenance is native.

Every decision points to the decisions that authorized it. Authority is not inferred after the fact; it is embedded at creation time.

Audits become graph traversals rather than investigations.

TEMPORAL CONSISTENCY

Because decisions are time-bound, the fabric encodes temporal truth.

It becomes possible to ask precise questions:

> What authority existed at this moment?

> Which decisions were valid when this action occurred?

> What changed between two points in time?

These questions are difficult or impossible to answer reliably in log-centric systems.

COMPOSABILITY OF DECISION TYPES

Binary approvals, multi-choice votes, role grants, and policy exceptions are often implemented as separate systems.

In a decision fabric, they are variations of the same primitive.

This composability reduces conceptual fragmentation. Governance logic becomes consistent across domains that were previously siloed.

EXECUTION IS A CONSUMER

In traditional architectures, *execution drives authority*.

In a decision fabric, **execution consumes authority**.

Systems act only when a valid decision exists that authorizes the action within scope and time. Automation becomes safer because it is downstream of explicit human commitment.

FAILURE BECOMES CONTAINED

When authority is implicit, *failures cascade*.

When authority is explicit, **failures are bounded**.

A compromised credential cannot exceed the scope of its decisions. An automation bug cannot invent authority. Damage is limited by design.

HUMAN AND MACHINE DECISIONS

Not all decisions must be made by humans.

Machines can propose, recommend, and even decide, but only within authority that was explicitly granted. The fabric distinguishes between *who decided* and *who executed*.

This distinction is critical in AI-driven systems.

OBSERVABILITY WITHOUT SURVEILLANCE

The decision fabric is observable without being invasive.

Because decisions are explicit artifacts, systems can monitor governance health without inspecting behavior. Oversight focuses on authority flows, not personal activity.

Privacy improves as accountability increases.

FROM FABRIC TO INFRASTRUCTURE

Once decisions are linked, verifiable, and bounded, they form infrastructure.

This infrastructure is reusable across domains: security, finance, compliance, operations, and governance.

Trust stops being an emergent property and becomes an engineered one.

WHAT THIS ENABLES

A decision fabric enables:

- unified governance models
- deterministic audits
- principled automation
- privacy-preserving oversight

It is the foundation on which systems like Privacy First are built.

————

In the next chapter, we shift from model to mechanism, examining how modern authentication, passkeys, and cryptographic standards make the decision fabric practical at scale.

PART IV

GOVERNANCE IN THE
AGE OF AI AND PRIVACY

Authentication has always been asked to do too much.

It was expected to establish identity, prevent intrusion, signal intent, and justify authority; all while remaining invisible to users and inexpensive to operate.

It failed not because of poor implementation, but because it was never designed to support governance.

Rules describe intent; decisions create obligation.

CHAPTER 12
AUTHENTICATION REVISITED

WHY PASSKEYS CHANGE EVERYTHING

Passwords are shared secrets.

They can be copied, replayed, phished, leaked, guessed, and reused. None of these failures are edge cases; they are structural properties of the model.

More importantly, passwords establish no presence.

A password can be used by anyone, anywhere, at any time, without the knowledge or awareness of its owner. The system cannot distinguish deliberate action from silent misuse.

Passwords authenticate accounts. They cannot authenticate intent.

MFA DID NOT FIX THE MODEL

Multi-factor authentication reduced certain classes of attack, but it did not change the underlying assumption.

Factors are still credentials. They are still reusable. They still outlive attention. Once verified, authority persists until explicitly revoked.

MFA hardened the perimeter. It did not create proof.

AUTHENTICATION WITHOUT PRESENCE

Most authentication systems optimize for session longevity.

Once authenticated, actors remain authorized long after the moment of verification. Decisions made minutes or hours later inherit authority by proximity, not by proof.

This gap is where legitimacy erodes.

PASSKEYS CHANGE THE QUESTION

Passkeys are not better passwords.

They change the question authentication answers.

Instead of asking:

> Does this actor know a secret?

Passkeys ask:

> Is this actor present, now, and willing to sign?

This shift is foundational.

CRYPTOGRAPHIC PRESENCE BY DESIGN

Passkeys are based on asymmetric cryptography.

The private key never leaves the device. Authentication requires a live cryptographic operation bound to a challenge issued in real time.

There is no reusable secret. There is no silent replay.

Presence is enforced by protocol.

USER VERIFICATION AS A REQUIREMENT

Modern passkey systems require user verification.

A biometric gesture or local PIN is not a factor in the traditional sense. It is a gate on key use.

The system does not trust the biometric. It trusts the signature that could only have been produced after verification.

This distinction matters.

WHY THIS ENABLES DECISIONS

A system that can require presence and produce signatures on demand can support explicit decisions.

Every critical action can be paired with:

- a unique challenge
- an explicit acknowledgment
- a cryptographic commitment

Authentication becomes an ingredient in governance rather than a substitute for it.

FROM SESSIONS TO MOMENTS

Passkeys favor moment-based authority.

Instead of extending trust across time, they make it cheap to reassert presence exactly when authority is exercised.

This aligns authentication with how legitimacy actually works.

RESISTANCE TO PHISHING AND REPLAY

Because passkeys are origin-bound and challenge-specific, they are unphishable by design.

This is not an incremental improvement. It eliminates entire categories of attack that undermine decision legitimacy.

When proof cannot be faked, trust can be structural.

AUTHENTICATION REFRAMED

With passkeys, authentication stops being the first step in a long chain of assumed authority.

It becomes a reusable capability: the ability to prove presence and intent whenever it matters.

This reframing is what makes the decision fabric practical at scale.

WHY THIS MATTERS BEYOND LOGIN

When authentication is moment-bound and cryptographically strong, it can be invoked for:

- approvals
- votes
- grants
- high-impact actions

Login becomes just one use case among many.

A NECESSARY, NOT SUFFICIENT, CONDITION

Passkeys do not create governance on their own.

They make it possible.

Without explicit decisions, bounded authority, and a decision fabric, passkeys would simply be a better lock on the same old door.

With them, they become the keystone of cryptographic governance.

———

In the next chapter, we examine how zero-knowledge proofs, selective disclosure, and decentralized identity extend the decision fabric without sacrificing privacy.

CHAPTER 13
PRIVACY WITHOUT BLINDNESS
PROOF WITHOUT EXPOSURE

Privacy and proof are often treated as opposites.

The assumption is simple: the more a system can verify, the more it must see. Identity must be revealed to establish trust. Data must be collected to enforce accountability. Oversight requires visibility.

This assumption is wrong.

THE FALSE TRADEOFF

Most privacy failures are justified in the name of assurance.

Systems collect more data than necessary because they do not know how to prove less. Verification becomes synonymous with disclosure. Trust becomes synonymous with surveillance.

The result is a brittle equilibrium: systems are trusted only so long as they are not abused.

PROOF DOES NOT REQUIRE EXPOSURE

Cryptography breaks the link between verification and revelation.

It makes it possible to prove statements about data without revealing the data itself. To verify eligibility without learning identity. To confirm authority without exposing activity.

This is not theoretical. These techniques are already standardized and deployed.

ZERO-KNOWLEDGE PROOFS AS A GOVERNANCE PRIMITIVE

A zero-knowledge proof allows an actor to prove that a statement is true without revealing why it is true.

In governance terms, this means:

- proving qualification without disclosing attributes
- proving authorization without revealing identity
- proving compliance without exposing records

Legitimacy can be established *without expanding the attack surface*.

SELECTIVE DISCLOSURE AND CONTEXTUAL TRUTH

Not all decisions require full context.

Selective disclosure allows an actor to reveal only the claims necessary for a specific decision, and nothing more.

A system does not need to know *who* you are if all it needs to know is *that you are allowed*.

Truth becomes contextual rather than absolute.

CREDENTIALS WITHOUT CENTRALIZATION

Decentralized identity systems invert the traditional model.

Instead of organizations issuing accounts and storing identity data, individuals hold credentials and present proofs when needed.

Authority shifts away from directories and toward decisions.

This reduces correlation, limits data accumulation, and restores agency to the individual.

PREVENTING THE GOVERNANCE–SURVEILLANCE COLLAPSE

Governance systems that cannot minimize data eventually become surveillance systems.

Logs expand. Context accumulates. Oversight turns invasive.

By designing for proof without exposure, systems can increase accountability while decreasing observation.

This is not a compromise. It is an architectural choice.

AUDITS WITHOUT RAW DATA

When decisions are explicit and signed, audits no longer require exhaustive data access.

Auditors verify proofs, not behavior. They validate authority chains, not personal activity.

Compliance shifts from inspection to verification.

AI CHANGES THE PRIVACY EQUATION

AI systems thrive on correlation.

If governance systems centralize identity and decision data, AI will inevitably reconstruct behavior, relationships, and intent beyond what was authorized.

Privacy-preserving proofs limit what AI systems can infer, even when they are powerful.

PRIVACY AS A STRUCTURAL OUTCOME

Privacy cannot be enforced through policy alone.

It must emerge from design constraints that make over-collection unnecessary and overreach impossible.

When systems are built around explicit decisions and minimal proofs, *privacy is the default*.

SEEING LESS, KNOWING MORE

The goal of privacy-first governance is not blindness.

It is precision.

Systems should know exactly what they need to know, and nothing else.

TRUST WITHOUT EXPOSURE

When proof replaces observation, trust stops depending on good intentions.

It becomes mechanical.

This is the only trust model that scales in an AI-driven world.

———

In the next chapter, we examine how AI systems interact with the decision fabric , and why humans must remain the source of authority even as machines accelerate execution.

CHAPTER 14
HUMANS IN THE LOOP
AUTHORITY IN AN AI-DRIVEN SYSTEM

AI systems excel at speed, scale, and pattern recognition.

They do not possess authority.

As automation accelerates, this distinction becomes easy to blur. Systems recommend actions, trigger workflows, and execute changes with minimal human involvement. Over time, recommendation quietly becomes delegation.

This chapter draws a hard line.

AUTOMATION IS NOT AUTHORITY

Automation executes.

It follows instructions, applies models, and optimizes outcomes according to defined objectives. Even when systems appear autonomous, their power is always derivative.

Authority originates elsewhere.

Confusing execution with authority allows machines to act without legitimacy. The system may function correctly while governance quietly fails.

THE PROPOSAL-DECISION BOUNDARY

AI systems are well suited to propose actions:

- flag anomalies
- rank options
- simulate outcomes
- recommend responses

These proposals can be sophisticated, probabilistic, and adaptive.

Decisions are different.

A decision commits authority, acknowledges consequence, and accepts responsibility. These properties cannot be inferred from model output.

WHY HUMANS MUST DECIDE

Authority is inseparable from accountability.

Only humans can accept responsibility for outcomes that affect rights, resources, and obligations. Only humans can weigh context that lies outside data.

Removing humans from the decision loop does not eliminate responsibility. It obscures it.

PRESENCE CANNOT BE AUTOMATED

Earlier chapters established presence as a prerequisite for intent.

AI systems do not have presence. They cannot be aware of consequence, nor can they consent to it. They operate continuously, without attention or acknowledgment.

Authority exercised without presence is indistinguishable from error.

CONSTRAINING AI WITH MANDATES

AI systems should operate within mandates.

A mandate defines:

- what the system may propose
- what it may execute automatically
- where human approval is required

Within these bounds, AI can move fast without exceeding legitimacy.

HUMAN-IN-THE-LOOP BY DESIGN

Human oversight is often added as a checkpoint.

This approach is brittle. Under pressure, checkpoints are bypassed or rubber-stamped. The system optimizes around them.

When decisions are explicit and signed, human involvement becomes structural. Authority cannot be exercised without deliberate presence.

SCALING JUDGMENT WITHOUT DILUTION

The goal is not to slow systems down.

It is to scale judgment without diluting responsibility. AI accelerates analysis and execution. Humans retain authority over commitment.

This division of labor preserves both efficiency and legitimacy.

FAILURE MODES IN AI GOVERNANCE

When AI systems are allowed to decide:

- errors propagate faster
- accountability diffuses
- bias becomes harder to contest
- rollback becomes political rather than technical

Explicit decision boundaries prevent these failures from compounding.

TRUSTWORTHY ACCELERATION

AI should make systems faster, not riskier.

When authority is explicit, bounded, and human-signed, acceleration does not undermine trust. *It reinforces it*.

Speed becomes safe because it is constrained.

DESIGNING FOR THE INEVITABLE

AI capabilities will continue to improve.

The temptation to delegate authority will grow. Systems that cannot resist this pressure will eventually lose legitimacy.

Designing for human authority is not a temporary precaution. **It is a permanent requirement**.

THE ROLE OF INFRASTRUCTURE

Human-in-the-loop governance does not scale through vigilance.

It scales through infrastructure that makes improper delegation impossible.

The decision fabric provides this infrastructure.

———

In the next chapter, we move from principles to platform, examining how Privacy First operationalizes the decision fabric as a service.

PART V

THE PLATFORM

A model explains how something *should* work.

Infrastructure determines what *can* work.

Up to this point, we have described a way of thinking about authority, decisions, privacy, and trust. The temptation now is to treat these ideas as guidelines, principles to be implemented piecemeal within existing systems.

That approach fails.

Infrastructure reveals what values were assumed.

CHAPTER 15
FROM MODEL TO INFRASTRUCTURE
WHY GOVERNANCE MUST BE BUILT, NOT LAYERED

Most governance failures are not caused by missing features, but by misplaced responsibility.

When legitimacy depends on correct configuration, disciplined process, or human vigilance, it will eventually fail. Layers can be bypassed, misapplied, or quietly disabled. Exceptions accumulate. Shortcuts become permanent.

Governance that matters cannot be optional.

THE DIFFERENCE BETWEEN TOOLS AND INFRASTRUCTURE

Tools assist behavior.

Infrastructure constrains it.

A tool can help an organization make better decisions. Infrastructure determines which decisions are even possible. This distinction is subtle but decisive.

Logs, policies, workflows, and reviews are tools. They operate after authority has already been exercised.

Infrastructure operates before.

WHEN AUTHORITY IS AN EMERGENT PROPERTY

In most systems, authority emerges indirectly.

Permissions combine with roles. Sessions combine with policy. Outcomes are interpreted as intent. Legitimacy is inferred from behavior rather than proven at commitment.

This emergence is fragile. It depends on everything going right, everywhere, all the time.

Infrastructure cannot depend on hope.

ENCODING LEGITIMACY INTO THE SUBSTRATE

The decision fabric changes where legitimacy lives.

Instead of being reconstructed after the fact, legitimacy is encoded at the moment authority is exercised. Decisions are explicit. Intent is signed. Scope and time are bounded.

When these properties are part of the substrate, systems cannot accidentally exceed authority. They must be deliberately extended.

WHY EXISTING IAM SYSTEMS FALL SHORT

Traditional IAM platforms evolved to manage access, not authority.

They excel at provisioning accounts, evaluating policies, and enforcing permissions. They were never designed to model decisions, capture intent, or prove legitimacy.

Adding these capabilities as extensions does not change the foundation. It creates parallel systems of truth.

Governance fractures.

INFRASTRUCTURE AS A SHARED SERVICE

Legitimacy should not be reimplemented in every application.

Just as organizations rely on shared infrastructure for networking, storage, and compute, they require shared infrastructure for trust. Decisions, proofs, and authority boundaries must be consistent across domains.

This consistency cannot be achieved through convention alone.

FROM THEORY TO SERVICE

Operationalizing the decision fabric requires a system that:

- captures intent at the moment of decision
- binds authority cryptographically
- enforces scope and time by design
- minimizes data exposure
- integrates with existing execution systems

This system must be reliable, neutral, and boring.

Trust infrastructure should disappear into the background, and become painfully obvious only when it is missing.

INTRODUCING PRIVACY FIRST

Privacy First is an attempt to build this infrastructure.

It is not an application layer, a workflow engine, or a compliance dashboard. It does not replace business logic or organizational judgment.

It provides a shared substrate for making, proving, and enforcing legitimate decisions.

CONSTRAINT AS A FEATURE

Privacy First is defined as much by what it refuses to do as by what it enables.

It does not assume standing authority. It does not centralize identity unnecessarily. It does not observe behavior it does not need to see.

These constraints are not limitations. They are the conditions under which trust can exist.

INFRASTRUCTURE CHANGES INCENTIVES

When legitimacy is structural, incentives shift.

Shortcuts become harder than correctness. Excess privilege becomes visible. Automation accelerates safely because authority cannot drift silently.

The system rewards good governance by making it the path of least resistance.

A NECESSARY TRANSITION

Moving from model to infrastructure is not an optimization.

It is the difference between ideas that persuade and systems that endure.

———

In the next chapter, we define what Privacy First is, not as a product category, but as a system of responsibility, and clarify the architectural boundaries that make cryptographic governance possible.

CHAPTER 16
PRIVACY FIRST
WHAT KIND OF SYSTEM THIS IS

Most systems are defined by what they do.

Trust infrastructure must be defined by what it is responsible for, and what it is not.

PrivacyFirst.id does not fit cleanly into an existing product category. It resembles IAM, but does not behave like traditional IAM. It participates in workflows, but does not own business logic. It supports compliance, but does not generate narratives.

This ambiguity is intentional.

NOT AN APPLICATION LAYER

Privacy First is not where work happens.

It does not manage projects, move money, approve expenses, or execute policies. It does not contain organizational context beyond what is strictly necessary to prove authority.

Execution systems remain sovereign.

Privacy First exists only at the moment where authority must be established, constrained, and proven.

NOT A POLICY ENGINE

Policies describe conditions.

They do not express intent.

Privacy First does not attempt to infer whether a policy *should* apply. It records when a human (or a bounded automation) explicitly authorizes an outcome.

Policy engines may consult mandates. They do not replace them.

NOT A WORKFLOW ORCHESTRATION TOOL

Workflows optimize sequence.

They do not create legitimacy.

Privacy First does not decide who is next in a process, how tasks are routed, or when exceptions occur. It ensures that when a decision is required, that decision is explicit, signed, and bounded.

Everything else is orchestration.

A DECISION INFRASTRUCTURE

Privacy First is responsible for one thing:

establishing cryptographic proof that a specific actor authorized a specific outcome under explicit constraints.

This responsibility is narrow by design.

Within this boundary, the platform provides:

- identity presence through modern authentication
- decision capture through mandates
- authority materialization through grants
- immutable, verifiable records of commitment

Outside this boundary, it is deliberately silent.

AUTHORITY WITHOUT CENTRALIZATION

Privacy First does not seek to become an identity hub.

Identities may be federated, decentralized, or external. Credentials may be issued elsewhere. Attributes may remain with their source.

The platform's role is not to *know* users deeply, but to know — with certainty — when a real actor has exercised authority.

Less data produces stronger guarantees.

MINIMAL DATA AS A SECURITY PROPERTY

Data minimization is not a compliance posture.

It is a security strategy.

Privacy First retains only what is required to:

- verify signatures
- validate scope and time
- prove provenance

It does not retain behavioral exhaust, inferred intent, or secondary analytics.

What is not collected cannot be leaked, subpoenaed, or misused.

NEUTRAL BY CONSTRUCTION

Trust infrastructure must not have opinions about outcomes.

Privacy First does not evaluate whether a decision was *wise*, only whether it was *legitimate*. It does not optimize for business success, only for correctness of authority.

This neutrality is what allows the platform to be shared across domains without becoming a point of contention.

COMPOSABILITY OVER CONTROL

Privacy First is designed to be composed, not obeyed.

Applications, services, and automations can reference mandates, verify grants, and validate authority without coupling to internal platform state.

Cryptographic proofs travel.

This is how legitimacy scales without central control.

INFRASTRUCTURE, NOT INTERFACE

Privacy First has interfaces, but it is not an interface.

Its success is measured not by daily active users, but by the absence of ambiguity during audits, incidents, and disputes.

When authority is questioned, the system answers decisively, without interpretation.

A CATEGORY OF ONE

If Privacy First appears difficult to categorize, that is because the category itself is missing.

We have tools for identity, access, policy, and compliance. We do not have infrastructure for provable authority.

This system exists to fill that gap.

———

In the next chapter, we examine how authentication becomes proof of presence rather than a login ritual, and why passkeys are the enabling mechanism for decision-centric governance.

CHAPTER 17
AUTHENTICATION AS PRESENCE, NOT LOGIN

Most authentication systems answer the wrong question.

They ask:

> Who is this?

What governance requires is different:

> Is this person present now, and are they deliberately authorizing this?

Privacy First treats authentication not as a gate at the edge of a session, but as a proof of presence at the moment authority is exercised.

THE SESSION IS THE PROBLEM

Sessions were designed for convenience, not legitimacy.

They assume continuity of intent across time. They persist authority beyond attention. They allow actions to be performed long after the human has disengaged.

From a governance perspective, sessions are standing delegations of power.

Standing authority is indistinguishable from abdication.

LOGIN AS RITUAL

In most systems, login is ceremonial.

A user authenticates once, early, often without context. The system records identity. Authority is implied rather than proven.

Everything that follows is treated as authorized by default.

This model optimizes throughput, not legitimacy.

PRESENCE IS MOMENT-BOUND

Authority exists only at the moment it is exercised.

Privacy First therefore requires authentication to occur at the moment a decision is made, not at the beginning of a session and not inferred from recent activity.

Authentication becomes episodic, intentional, and contextual.

PASSKEYS AS PRESENCE PROOF

Passkeys change authentication because they change *what is being proven*.

A passkey operation is:

- challenge–response based
- cryptographically bound to origin
- resistant to replay
- resistant to delegation
- backed by user verification

Most importantly, it requires the user to be present.

This makes passkeys suitable not just for access, but for authority.

WHY PASSWORDS AND OTPS FAIL HERE

Passwords can be shared, stolen, replayed, or automated.

One-time codes reduce risk, but they do not prove presence. They can be proxied, relayed, or harvested. They authenticate *a channel*, not a person.

None of these mechanisms produce a durable act of commitment.

AUTHENTICATION AS COMMITMENT

When a user completes a passkey operation tied to a specific decision payload, they are not merely authenticating.

They are signing intent.

The cryptographic signature binds:

- the actor
- the decision
- the context
- the moment

This is authentication elevated to commitment.

RE-AUTHENTICATION IS NOT FRICTION

In a decision-centric system, re-authentication is not a nuisance.

It is the mechanism by which authority is renewed.

Privacy First requires re-authentication whenever a mandate is created, modified, or exercised. Authority does not carry forward implicitly.

This makes legitimacy explicit, and visible.

AUTOMATION WITHOUT STANDING POWER

Automation still operates.

But it operates within the bounds of mandates that were explicitly authorized. When automation reaches the edge of its authority, it stops.

At that boundary, presence is required again.

This is how systems scale without drifting into silent autonomy.

AUTHENTICATION THAT LEAVES EVIDENCE

Traditional authentication leaves logs.

Presence-based authentication leaves proof.

A signed authentication event can be verified independently, without trusting the platform that recorded it. It can be replayed for audit without being reused for action.

This distinction matters when decisions are challenged.

FEWER LOGINS, STRONGER AUTHORITY

Presence-based systems often reduce overall authentication events.

Users are not repeatedly challenged for low-risk activity. They are only required to authenticate when authority is actually exercised.

The result is less friction, and far stronger guarantees.

AUTHENTICATION REFRAMED

In Privacy First, authentication is not about access.

It is about legitimacy.

Once authentication is treated as proof of presence and commitment, the boundary between security and governance disappears.

––––––

In the next chapter, we examine how mandates, approvals, and grants emerge as different expressions of the same decision primitive; and how organizations model real authority without fragmenting their systems.

CHAPTER 18
MANDATES, APPROVALS, AND GRANTS

Organizations recognize authority through familiar patterns.

Someone approves an expense. A committee votes. A role is granted. Access is revoked.

These patterns appear different on the surface, but they share a common flaw in most systems: authority is implied by process rather than proven by decision.

Privacy First treats these patterns as variations of a single primitive.

ONE DECISION, MANY FORMS

At their core, approvals, votes, and role assignments are all answers to the same question:

> Has a legitimate actor authorized this outcome under defined constraints?

The difference lies not in authority, but in expression.

By modeling each as a mandate, Privacy First collapses complexity without reducing nuance.

APPROVALS AS BOUNDED MANDATES

An approval is the simplest expression of a mandate.

It presents a binary choice, approve or reject, within a defined scope and time. The actor's intent is explicit. The outcome is unambiguous.

What matters is not the simplicity of the choice, but the proof of commitment.

When an approval is signed, legitimacy is established at that moment, not inferred later from logs or policy.

MANDATES BEYOND BINARY CHOICE

Real governance is rarely binary.

Boards elect leaders. Teams prioritize initiatives. Organizations route decisions based on context and consequence.

Mandates allow choice sets of arbitrary size and structure. They can represent ranked options, multi-selection ballots, or conditional paths.

The decision remains explicit. The authority remains bounded.

GRANTS AS MATERIALIZED AUTHORITY

Roles do not create authority.

They express it.

A grant is the materialization of a mandate into executable capability. It translates a signed decision into permissions that other systems can enforce.

The grant inherits its legitimacy from the mandate that created it, including scope, time, and conditions.

REVOCATION AS DECISION, NOT CLEANUP

Revocation is often treated as administrative hygiene.

In a decision-centric system, revocation is itself a mandate. Authority is removed deliberately, with the same proof and constraints as when it was granted.

This symmetry matters. It ensures that loss of power is as explicit as its acquisition.

AVOIDING FRAGMENTED GOVERNANCE

Traditional systems separate approvals, voting, and access management into distinct tools.

Each develops its own logs, semantics, and failure modes. Authority fragments. Audits reconstruct across systems.

By unifying these expressions, Privacy First ensures that all authority decisions share the same guarantees, regardless of where they are exercised.

EXECUTION REMAINS EXTERNAL

Privacy First does not enforce outcomes.

Execution systems consume grants and verify mandates, but they retain control over business logic. This separation preserves flexibility while preventing silent authority.

Decisions are proven once. Execution happens many times.

COMPOSABILITY ACROSS DOMAINS

Because mandates are cryptographically verifiable artifacts, they can be referenced across systems without tight coupling.

A financial system, an HR platform, and an infrastructure stack can all rely on the same proof of authority, without sharing internal state.

This is how governance scales without centralization.

FAMILIAR PATTERNS, STRONGER GUARANTEES

Nothing about this model requires organizations to abandon familiar workflows.

What changes is what those workflows produce.

Instead of ephemeral approvals and implicit grants, they produce durable, verifiable authority.

AUTHORITY MADE LEGIBLE

When approvals, mandates, and grants are unified, authority becomes legible.

It can be inspected, reasoned about, and challenged without interpretation. Disputes resolve to proof, not memory.

This is governance that survives scale, turnover, and automation.

———

In the next chapter, we examine how this model produces auditability and compliance as a structural outcome, without expanding surveillance or data collection.

CHAPTER 19
AUDITABILITY WITHOUT SURVEILLANCE

Most audits fail for the same reason trust fails: evidence is reconstructed after the fact.

Logs are collected. Events are correlated. Narratives are assembled. Gaps are explained. Confidence is inferred.

This process is expensive, intrusive, and fragile; and it still cannot prove legitimacy.

Privacy First approaches auditability from the opposite direction.

WHY AUDITS BECOME THEATER

Traditional audits answer questions indirectly.

Auditors examine policies, configurations, access lists, and logs to infer whether decisions were made correctly. They look for consistency rather than proof.

When inconsistencies appear, organizations respond with explanations.

The result is compliance theater: convincing stories built on incomplete evidence.

EVIDENCE AT THE MOMENT OF COMMITMENT

A signed decision is evidence.

It does not require interpretation. It does not rely on system behavior or operator intent. It states, unambiguously, that a specific actor authorized a specific outcome under explicit constraints.

When evidence is created at commitment, audits stop being investigative.

They become mechanical.

DETERMINISTIC AUDITS

In a decision-centric system, an audit becomes a verification exercise.

Each question resolves to:

- does a mandate exist?
- is the signature valid?
- were scope and time respected?

There is no need to reconstruct timelines or infer intent. Proof is self-contained.

This determinism reduces both cost and contention.

PROVENANCE WITHOUT SURVEILLANCE

Most compliance programs expand data collection in pursuit of certainty.

More logs. More monitoring. More behavioral analysis.

This creates privacy risk without delivering proof.

Privacy First achieves provenance without surveillance by retaining only cryptographic artifacts required to verify authority. Behavior remains outside the system.

PRIVACY IMPROVES AS EVIDENCE IMPROVES

This is counterintuitive but consistent.

When systems cannot prove legitimacy, they compensate by observing more. When legitimacy is explicit, observation becomes unnecessary.

Stronger evidence allows weaker monitoring.

Privacy becomes an outcome of better governance, not a constraint upon it.

COMPLIANCE BY CONSTRUCTION

Regulatory frameworks demand different things on the surface.

SOX focuses on financial controls. SOC 2 emphasizes process integrity. GDPR mandates minimization and accountability.

At their core, they all ask the same question:

> Can you prove that authority was exercised appropriately?

Decision-centric infrastructure answers this once — structurally.

AUDITS THAT DO NOT SCALE INCREASE RISK

Traditional audits increase risk.

They centralize sensitive data, expand access, and expose systems to third parties. The act of auditing becomes an attack surface.

When evidence is cryptographic and minimal, audits can be performed without broad access or data duplication.

Risk decreases as verification increases.

DISPUTES RESOLVE TO PROOF

When decisions are contested, organizations often fall back on memory, testimony, or policy interpretation.

These disputes are slow and political.

With signed mandates, disputes resolve to verification. Either authority existed within bounds, or it did not.

This clarity protects both organizations and individuals.

LONG-TERM VERIFIABILITY

Logs decay.

Systems change. Staff turnover. Policies evolve. Context is lost.

Cryptographic proof does not degrade with time. A decision signed years ago can be verified with the same certainty as on the day it was made.

This durability is essential for regulated environments.

AUDITABILITY AS INFRASTRUCTURE

When auditability is layered on, it becomes burdensome.

When it is infrastructural, it becomes invisible.

Privacy First treats auditability as a natural consequence of explicit authority, not as a reporting obligation.

COMPLIANCE WITHOUT COMPROMISE

Organizations should not have to choose between compliance and privacy.

When legitimacy is provable, both improve.

This is the difference between being compliant and being correct.

———

In the next chapter, we examine how Privacy First accommodates AI-driven analysis and automation without allowing authority to drift beyond human control.

CHAPTER 20
AI-READY
BY DESIGN

Automation promises speed.

Unchecked, it also dissolves authority.

The challenge is not whether AI systems should participate in decision-making processes, they already do. The challenge is ensuring that as systems accelerate, legitimacy does not erode.

Privacy First is designed for this reality.

THE ACCELERATION PARADOX

AI systems increase velocity.

They surface insights faster than humans, identify patterns at scale, and propose actions continuously. This creates pressure to collapse decision boundaries.

Under pressure, recommendation becomes execution.

Governance fails quietly.

PROPOSAL IS NOT COMMITMENT

AI excels at proposing.

It can rank options, assign risk scores, predict outcomes, and simulate consequences. These capabilities are valuable, but they are not decisions.

A decision commits authority.

Privacy First enforces this distinction structurally. AI may propose. Only mandates may commit.

BOUNDED AUTOMATION

Automation operates within limits.

Those limits are defined by mandates that specify scope, duration, and consequence. Within these bounds, systems can act freely and at speed.

When automation reaches the edge of its authority, it stops.

Presence is required again.

RISK SCORING WITHOUT DELEGATION

Risk engines are often used to silently adjust permissions or trigger actions.

This shifts authority into opaque systems.

In a decision-centric architecture, risk scores inform mandates. They do not replace them. Humans remain responsible for accepting or rejecting risk.

JUST-IN-TIME AUTHORITY

AI systems thrive on context.

Privacy First supports just-in-time mandates that grant authority precisely when required, and expire automatically when context changes.

This minimizes standing power while preserving responsiveness.

EXPLAINABILITY THROUGH STRUCTURE

Explainability is often framed as a model property.

It is more reliably achieved as a governance property.

When AI-driven actions are bounded by explicit mandates, their legitimacy is explainable even if the underlying model is complex.

The question shifts from *why did the model act* to *who authorized this outcome, and under what constraints*.

FAILURE CONTAINMENT

When automation fails, impact matters more than cause.

By constraining authority, Privacy First ensures that failures remain local. Blast radius is limited by design.

Rollback becomes technical, not political.

ACCELERATING WITHOUT ABDICATING

The goal is not to slow systems down.

It is to accelerate without surrendering control. When

authority is explicit, automation becomes safer as it becomes faster.

This reverses the usual tradeoff between speed and trust.

AI AS A DEPENDENT ACTOR

In Privacy First, AI systems are dependent actors.

They operate only within authority that has been deliberately granted. They cannot accumulate power silently or expand scope opportunistically.

This dependency is structural, not procedural.

DESIGNING FOR THE INEVITABLE

AI capabilities will continue to improve.

Systems that rely on vigilance or policy to constrain them will eventually fail. Systems that encode authority into infrastructure will endure.

AI-readiness is not about model integration. It is about authority preservation.

SAFE ACCELERATION

When authority is provable, speed becomes safe.

Privacy First does not resist automation. It makes it legitimate.

———

In the next chapter, we make explicit what Privacy First refuses to do, and why those refusals are essential to trust.

CHAPTER 21
WHAT PRIVACY FIRST REFUSES TO DO

Trust is not established by capability alone.

It is established by constraint.

In systems that wield authority, what is *possible* matters less than what is *forbidden*. The strongest guarantees do not come from features, but from refusals that are structural and irreversible.

Privacy First is defined as much by what it refuses to do as by what it enables.

NO SILENT AUTHORITY

Privacy First refuses to allow authority to be exercised without explicit, signed intent.

There are no default approvals. No inherited sessions. No background elevation of privilege. Authority does not persist beyond presence.

If a decision matters, it must be deliberate.

NO STANDING POWER

Standing authority is convenient.

It is also dangerous.

Privacy First refuses to grant indefinite permissions, unbounded roles, or evergreen access. All authority is scoped, time-bound, and revocable by design.

Power that cannot expire cannot be governed.

NO BEHAVIORAL SURVEILLANCE

Privacy First refuses to observe what it does not need to know.

It does not profile users, infer intent from behavior, or collect exhaust for secondary analysis. Risk is evaluated at the moment of decision, not accumulated through observation.

Privacy is preserved by absence, not policy.

NO IDENTITY CENTRALIZATION

Privacy First refuses to become an identity hub.

It does not require exclusive control over identities, attributes, or credentials. Federation, decentralization, and external attestations are first-class citizens.

Authority is proven through signatures, not ownership of identity data.

NO OPAQUE AUTOMATION

Privacy First refuses to let machines decide in silence.

AI systems may recommend, rank, and propose. They may not commit authority without an explicit mandate. Automation does not accumulate power over time.

Opacity is incompatible with legitimacy.

NO NARRATIVE AUDITS

Privacy First refuses to generate stories in place of evidence.

It does not reconstruct intent from logs or infer legitimacy from configuration. Audits resolve to cryptographic verification, not interpretation.

Proof replaces persuasion.

NO DATA HOARDING

Privacy First refuses to collect data "just in case."

Retention is minimal, purpose-bound, and explicit. What cannot be justified for proof is not stored.

This refusal reduces both risk and temptation.

NO VENDOR LOCK-IN THROUGH CONTROL

Privacy First refuses to trap authority inside proprietary state.

Mandates and grants are verifiable artifacts. They can be inspected, validated, and relied upon without trusting the platform that issued them.

Legitimacy must outlive vendors.

NO CONFIGURATION AS GOVERNANCE

Privacy First refuses to treat configuration as a substitute for decision-making.

Policies and settings define possibilities. They do not create legitimacy. When authority is required, a human decision must occur.

Governance cannot be precomputed.

WHY THESE REFUSALS MATTER

Each refusal removes a class of failure.

Together, they ensure that authority cannot drift, accumulate, or disappear into abstraction. The system remains legible under pressure, audit, and scale.

These constraints are not defensive. They are generative.

TRUST THROUGH IMPOSSIBILITY

The highest form of trust is not confidence.

It is impossibility.

When systems make the wrong actions impossible rather than unlikely, trust becomes structural.

This is the role Privacy First exists to play.

———

In the next part, we step back from the platform to examine what changes when authority becomes explicit, for organizations, regulation, and the future of digital trust.

PART VI

WHEN AUTHORITY
BECOMES EXPLICIT

Most organizations run on power they cannot see.

Authority accumulates through tenure, proximity, access, and convention. Decisions are made because someone *can* make them, not because they were explicitly empowered to do so.

This works until it doesn't.

When authority is implicit, accountability is negotiable.

When proof is native, trust becomes optional.

CHAPTER 22
ORGANIZATIONS WITHOUT IMPLICIT POWER

Organizational charts describe reporting lines.

They do not describe authority.

Real power flows through email inboxes, shared folders, standing permissions, and informal expectations. Over time, these pathways harden into precedent.

No single decision grants this power. It emerges.

Emergent authority is impossible to govern.

WHEN AUTHORITY MUST BE SHOWN

In a decision-centric system, authority cannot hide.

Every consequential action resolves to a mandate. If no mandate exists, authority does not exist. Informal influence may still operate, but it cannot produce binding outcomes.

This forces organizations to confront a simple question:

> Who is actually allowed to decide this?

FEWER ROLES, MORE DECISIONS

Traditional organizations attempt to manage authority by defining roles.

Roles scale poorly. They become broad, permanent, and over-privileged. Exceptions accumulate. Temporary needs become standing power.

Explicit decision systems invert this model.

Roles shrink. Decisions multiply.

Authority is granted where it is needed, when it is needed, and nowhere else.

ACCOUNTABILITY WITHOUT SURVEILLANCE

When authority is implicit, organizations compensate with monitoring.

They log activity, review behavior, and investigate outcomes. Accountability becomes retrospective and adversarial.

When authority is explicit, accountability becomes immediate. Responsibility is visible at the moment of commitment. Surveillance becomes unnecessary.

This shifts culture as much as control.

REDUCED BLAST RADIUS

Implicit authority fails catastrophically.

A compromised account, a mistaken script, or an overconfident employee can act far beyond what was intended.

Bounded mandates localize failure. Even when mistakes occur, their impact is constrained by scope and time.

Resilience improves without slowing execution.

DECISION FRICTION WHERE IT MATTERS

Not all friction is waste.

Explicit authority introduces deliberation at points of consequence, not everywhere. Low-risk activity remains fluid. High-risk decisions require presence and intent.

Organizations become both faster and safer, a combination that implicit power cannot achieve.

POWER THAT CAN BE CHALLENGED

Implicit authority is difficult to contest.

It relies on norms, assumptions, and seniority. Challenging it feels political.

Explicit authority is technical. Either a mandate exists, or it does not. Disputes resolve to proof rather than position.

This protects individuals as much as institutions.

CULTURAL RESISTANCE

Making authority explicit feels threatening.

It exposes informal power structures. It removes ambiguity that some rely on. Resistance is natural.

Organizations that endure this transition gain clarity. Those that do not remain fragile.

FROM HIERARCHY TO TOPOLOGY

When decisions are explicit, organizations can reason about authority structurally.

Authority becomes a graph: nodes of responsibility connected by bounded decisions. Hierarchy remains, but it no longer carries unchecked power.

This topology adapts better to scale, change, and automation.

THE END OF "BECAUSE I COULD"

Implicit power allows decisions to be justified after the fact.

Explicit authority requires justification in advance.

This single shift changes how organizations behave under pressure.

LEGITIMACY AS AN INTERNAL PROPERTY

Organizations often focus on external legitimacy: regulators, customers, markets.

Internal legitimacy matters first. When employees can see how authority is granted and exercised, trust increases.

Clarity replaces rumor. Proof replaces politics.

ORGANIZATIONS THAT CAN EXPLAIN THEMSELVES

When authority is explicit, organizations can explain their actions without defensiveness.

They do not rely on narratives or exceptions. They show decisions.

This capability will matter increasingly as organizations operate under greater scrutiny.

———

In the next chapter, we examine how regulation changes when oversight gives way to verification, and why explicit authority benefits regulators as much as organizations.

CHAPTER 23
REGULATION AS VERIFICATION, NOT OVERSIGHT

Regulation is often framed as a burden imposed on organizations.

In reality, it is a response to uncertainty.

When authority cannot be proven, regulators compensate by supervising behavior. They mandate controls, require reporting, and expand oversight. Complexity grows because evidence is weak.

Explicit authority changes this dynamic.

OVERSIGHT AS A SUBSTITUTE FOR PROOF

Traditional regulation assumes that organizations cannot reliably demonstrate legitimacy on their own.

As a result, regulators intervene upstream. They prescribe processes, dictate controls, and require continuous reporting. Compliance becomes a proxy for trust.

This model scales poorly.

VERIFICATION CHANGES THE ROLE OF THE REGULATOR

When decisions are explicit and cryptographically provable, the regulator's task simplifies.

They no longer need to monitor how systems are configured or how employees behave day to day. They verify whether authority existed when it mattered.

Regulation shifts from supervision to verification.

EVIDENCE THAT TRAVELS

Regulatory regimes are fragmented by jurisdiction.

Evidence is not.

A signed decision can be verified anywhere, by anyone with the appropriate mandate, without access to internal systems. Proof crosses borders more easily than process.

This property matters as organizations operate globally.

REDUCING REPORTING WITHOUT REDUCING ACCOUNTABILITY

Reporting requirements expand when trust contracts.

Verification allows reporting to shrink without weakening accountability. Regulators request fewer artifacts because each artifact carries stronger guarantees.

This reduces cost for organizations and noise for regulators.

LESS DISCRETION, MORE FAIRNESS

Oversight introduces discretion.

Inspectors interpret controls. Auditors assess intent. Outcomes vary based on narrative skill as much as behavior.

Verification reduces subjectivity. Either authority was exercised within bounds, or it was not. Enforcement becomes more consistent.

REGULATION THAT SURVIVES AUTOMATION

AI-driven systems strain traditional regulatory models.

Continuous oversight cannot keep pace with automated decision-making. Manual reviews lag behind machine speed.

Verification scales with automation because proof is generated at the moment of commitment.

Regulation remains effective even as systems accelerate.

INCENTIVES REALIGNED

When regulators verify decisions rather than supervise processes, organizational incentives change.

Firms invest in correctness rather than appearance. Engineering replaces paperwork. Governance becomes an asset rather than overhead.

This alignment benefits both sides.

MINIMAL EXPOSURE FOR REGULATORS

Oversight requires access.

Verification requires proof.

By relying on cryptographic evidence, regulators reduce their need to ingest sensitive data. Privacy improves for individuals and liability decreases for agencies.

This is governance with less intrusion.

FROM CHECKLISTS TO CLAIMS

Compliance checklists enumerate controls.

Decision-centric regulation evaluates claims:

- this decision was authorized
- by this actor
- within these bounds

Claims can be verified directly.

ENFORCEMENT BECOMES MECHANICAL

When proof is explicit, enforcement stops being adversarial.

Disputes resolve to verification rather than negotiation. Appeals focus on scope and mandate, not narrative reconstruction.

The rule of law becomes technical.

REGULATION AS A CONSUMER OF PROOF

In this model, regulators are not designers of systems.

They are consumers of evidence.

This separation preserves institutional independence while raising the quality of enforcement.

TOWARD TRUSTABLE REGULATION

Regulation exists to protect the public.

When authority is explicit, that protection becomes more precise and less invasive.

Verification replaces oversight. Proof replaces process.

———

In the next chapter, we examine how privacy changes when systems stop observing behavior and start requiring explicit consent and commitment.

CHAPTER 24
PRIVACY AS A
CIVIC PROPERTY

Privacy is often framed as a personal preference.

Something individuals trade for convenience, exchange for access, or surrender under pressure. In this framing, privacy is fragile, easily overridden by incentives, fear, or necessity.

This framing is incomplete.

Privacy is not only an individual concern. It is a civic property.

WHEN SYSTEMS CANNOT PROVE LEGITIMACY

Surveillance thrives where legitimacy is uncertain.

When systems cannot prove that authority was exercised correctly, they compensate by watching everything. Behavior is logged, analyzed, and retained in case it becomes relevant later.

Observation becomes the substitute for proof.

THE SOCIAL COST OF OBSERVATION

Pervasive monitoring does more than expose data.

It alters behavior. People adapt to being watched. They avoid risk, suppress dissent, and conform preemptively. Over time, surveillance reshapes culture.

This cost is diffuse, but real.

PRIVACY BEYOND SECRECY

Privacy is not about hiding wrongdoing.

It is about preserving the space in which legitimate action does not require justification in advance. It allows individuals to act, explore, and decide without constant external evaluation.

This space is essential for democratic participation and organizational trust.

CONSENT AS A CRYPTOGRAPHIC ACT

In most systems, consent is implied.

A checkbox is clicked. A policy is accepted. Authority is assumed.

In a decision-centric system, consent is explicit. It is bound to a specific outcome, at a specific moment, and proven cryptographically.

Consent becomes an act, not an assumption.

SELECTIVE DISCLOSURE AND DIGNITY

When proof requires full exposure, privacy erodes.

Modern cryptographic techniques: selective disclosure, zero-knowledge proofs, decentralized credentials; allow claims to be verified without revealing underlying data.

Dignity is preserved because individuals reveal only what is necessary.

FEWER REASONS TO WATCH

Explicit authority changes incentives.

When decisions are proven, systems no longer need to infer intent from behavior. Monitoring becomes unnecessary for legitimacy. Observation recedes to operational necessity.

Privacy improves because surveillance loses its justification.

PRIVACY AS A COLLECTIVE OUTCOME

Individual privacy protections fail when systems are designed to observe.

Civic privacy emerges when systems are designed to decide.

This distinction matters. One relies on restraint. The other relies on structure.

TRUST WITHOUT EXPOSURE

When authority is explicit, trust no longer depends on visibility into behavior.

Institutions can be held accountable without inspecting everyone. Individuals can participate without being profiled.

Trust becomes compatible with scale.

RESISTING THE DRIFT TOWARD TOTAL RECALL

Data storage is cheap.

The temptation to retain everything is constant. Once stored, data attracts new uses. Purpose limitation erodes quietly.

Decision-centric systems resist this drift by design. If data is not required for proof, it is not retained.

PRIVACY AS INFRASTRUCTURE

Privacy protections added after the fact are brittle.

When privacy is infrastructural, it is durable. It does not depend on policy updates or enforcement vigilance.

It depends on what the system cannot do.

A PUBLIC GOOD

Like clean air or safe roads, privacy benefits those who do not actively defend it.

When institutions minimize observation, everyone gains freedom of action. When they do not, everyone pays the cost.

Privacy is therefore not merely personal. It is civic.

RECLAIMING THE DEFAULT

The modern default is observation.

Decision-centric governance restores a different default: explicit consent at moments that matter, silence everywhere else.

This shift is subtle, and transformative.

———

In the next chapter, we examine how identity, authentication, and authority evolve beyond passwords, and what a post-credential world looks like.

CHAPTER 25
THE END OF THE PASSWORD ERA

Passwords were never meant to last.

They were a pragmatic solution to an early problem: how to recognize a returning user in a stateless system. Over time, that narrow mechanism was burdened with responsibility it was never designed to carry.

Authority, identity, and trust were all built on top of a shared secret.

The result has been fragility at scale.

SECRETS AS A FOUNDATION

Passwords treat identity as knowledge.

If you know the secret, you are the user. This model assumes secrecy, exclusivity, and careful handling, assumptions that fail under automation, phishing, reuse, and breach.

More controls were added. Complexity increased. Trust did not.

THE CREDENTIAL SPIRAL

As passwords weakened, systems layered compensations:

- password rotation
- complexity requirements
- one-time codes
- recovery questions

Each addition addressed symptoms, not structure. Credentials remained transferable, replayable, and divorced from presence.

Authority continued to rest on possession of secrets.

AUTHENTICATION WITHOUT SECRETS

Passkeys represent a break, not an improvement.

They eliminate shared secrets entirely. Authentication becomes a cryptographic exchange bound to a device, an origin, and a moment of user verification.

There is nothing to remember. Nothing to reuse. Nothing to steal remotely.

IDENTITY DECOUPLED FROM ACCOUNTS

Passwords bind identity to accounts.

In a post-credential world, identity becomes contextual. It is proven when required, for a specific purpose, without establishing a persistent account relationship.

Presence replaces possession.

FROM LOGIN TO PROOF

The login ceremony disappears.

What remains is proof at moments of consequence. Users are not authenticated to systems broadly. They prove presence and intent narrowly.

This aligns authentication with authority rather than access.

SELECTIVE IDENTITY

Traditional identity systems disclose more than they need.

Names, identifiers, attributes, and history are bundled together and shared wholesale. This creates unnecessary exposure.

Selective disclosure allows claims to be proven without revealing the underlying identity. Authority can be verified without making the user legible everywhere.

DECENTRALIZED CREDENTIALS

Decentralized identity shifts control.

Credentials can be issued by many parties and held by individuals. Verification does not require a central directory or global identifier.

Trust becomes a property of cryptography rather than registration.

AUTHORITY WITHOUT ACCOUNTS

When authentication is moment-bound and identity is selective, accounts lose their centrality.

Systems no longer need to know *who you are* in general. They need to know *that you are authorized for this decision now*.

This is a narrower, safer requirement.

FEWER HONEYPOTS

Credential databases are attractive targets.

Eliminating passwords removes entire classes of breach. There is no credential to exfiltrate, sell, or replay.

Risk shifts from centralized secrets to bounded, verifiable acts.

ECOSYSTEM EFFECTS

As credentials disappear, interoperability improves.

Users are no longer forced to manage dozens of identities. Systems no longer need to store sensitive secrets. Trust flows through proof rather than federation agreements.

This enables new forms of collaboration without central authority.

THE QUIET TRANSITION

The end of the password era will not feel dramatic.

There will be no single migration day. Systems will gradually stop asking users to prove who they are all the time.

They will only ask when it matters.

BEYOND AUTHENTICATION

The disappearance of passwords is not the end goal.

It is the precondition for decision-centric governance. Without secrets, authority can no longer be assumed. It must be proven.

This is the deeper shift underway.

———

In the next chapter, we examine how these primitives extend to autonomous and agentic systems, and how authority can be delegated without losing human control.

CHAPTER 26
DELEGATION WITHOUT ABDICATION

Delegation has always been a risk.

To delegate is to extend authority beyond direct control. It allows scale, speed, and specialization, but it also creates distance between decision and accountability.

As systems become autonomous and agentic, this risk intensifies.

AUTOMATION IS NOT AUTHORITY

Automation executes.

Authority decides.

Conflating the two leads to systems that act correctly until they do not, and then cannot explain why. When authority is implicit, automation inherits power accidentally.

This is abdication, not delegation.

THE AGENT PROBLEM

AI agents operate continuously.

They observe, infer, recommend, and increasingly act. Without explicit boundaries, they accumulate standing power, the ability to make decisions without renewed consent.

This is incompatible with human accountability.

BOUNDED AUTHORITY

Delegation must be explicit.

A mandate can grant an agent authority that is:

- scoped to specific actions
- limited in time
- constrained by context
- revocable without ambiguity

Anything less is trust by assumption.

MANDATES AS DELEGATION PRIMITIVES

In a decision-centric system, delegation is not a configuration.

It is a signed act.

A human grants authority to an agent through a mandate that specifies what the agent may do and under which conditions. The mandate is verifiable, auditable, and finite.

Authority becomes portable without becoming permanent.

PRESENCE AT THE POINT OF GRANT

Even if agents act autonomously, delegation does not.

The grant of authority requires human presence, intent, and proof. Once issued, the agent operates within those bounds, no more, no less.

Human agency is preserved where it matters.

REVOCATION AS A FIRST-CLASS ACT

Delegated authority must be retractable.

Revocation is not a policy update or configuration change. It is a signed decision that invalidates future actions.

This symmetry matters. Authority begins and ends explicitly.

AUDITABILITY WITHOUT SURVEILLANCE

Agent actions can be evaluated without continuous monitoring.

Each action either references a valid mandate or it does not. Compliance is determined by verification, not observation.

This preserves privacy while maintaining control.

PREVENTING AUTHORITY CREEP

Long-lived agents drift.

Scopes expand informally. Exceptions accumulate. Temporary permissions become permanent.

Mandates resist this drift by expiring, constraining, and requiring renewal. Authority remains intentional.

ACCOUNTABILITY REMAINS HUMAN

Agents do not bear responsibility.

Humans do.

By tying agent authority to explicit human decisions, accountability remains traceable. When something goes wrong, the question is not *what did the system do*, but *who authorized it*.

SCALING TRUST SAFELY

Delegation is necessary for scale.

Mandated delegation allows organizations to use autonomous systems without surrendering control. Trust scales through structure rather than optimism.

A FUTURE WITH AGENTS

Agentic systems are inevitable.

Unchecked authority is not.

By separating execution from authorization, and delegation from abdication, organizations can embrace autonomy without erasing human responsibility.

———

In the next chapter, we examine how these principles reshape digital ecosystems, and what a trust-native internet might look like.

CHAPTER 27
A TRUST-NATIVE INTERNET

The internet was built to move information.

Trust was added later.

As networks expanded, identity systems, platforms, and intermediaries emerged to compensate for the absence of native trust primitives. Authority was centralized because proof was scarce.

A different architecture is now possible.

TRUST AS AN OVERLAY

Most of today's trust lives above the network.

Platforms authenticate users. Institutions vouch for actions. Reputation systems infer legitimacy over time. Each layer compensates for the same missing capability: provable authority at the moment of action.

This produces fragility and concentration.

WHEN PROOF BECOMES NATIVE

A trust-native internet does not rely on intermediaries to assert legitimacy.

Actions carry their own proof. Decisions are verifiable wherever they are presented. Authority is explicit, portable, and bounded.

Trust becomes a property of artifacts, not platforms.

INTEROPERABILITY WITHOUT FEDERATION

Federation assumes shared governance.

Proof does not.

When decisions are cryptographically verifiable, systems interoperate without prior agreements. They verify claims directly rather than deferring to a central authority.

This lowers barriers to collaboration.

FEWER GATEKEEPERS

Gatekeepers exist where verification is hard.

When proof is easy, gatekeeping loses its justification. Power shifts away from those who control access toward those who produce legitimate decisions.

This redistribution is structural, not political.

MARKETS OF PROOF

In a trust-native environment, value flows with verification.

Services compete on the quality, clarity, and constraints of the authority they require and emit. Users choose systems that ask for less, prove more, and expire commitments cleanly.

Trust becomes composable.

GLOBAL BY DEFAULT

Proof ignores borders.

A signed decision can be verified without reference to jurisdiction, language, or local process. Regulatory interpretation may differ, but evidence does not.

This enables global interaction without global platforms.

REDUCED SURVEILLANCE INCENTIVES

When legitimacy is explicit, observation is unnecessary.

Systems no longer need to watch behavior to infer trust. They verify decisions directly. Surveillance becomes a cost without benefit.

Privacy improves as a side effect of better architecture.

RESILIENCE THROUGH DECENTRALIZATION

Centralized trust fails catastrophically.

When a platform is compromised, trust collapses for everyone. In a trust-native model, compromise is local. Proofs remain valid. Authority does not cascade.

Failure becomes contained.

COMPOSABILITY OF AUTHORITY

Explicit authority can be chained, nested, and constrained.

202 THE HYPERPROOF MANIFESTO

Mandates reference mandates. Decisions inherit bounds. Complex collaborations form without shared backends.

This enables ecosystems rather than monopolies.

CULTURAL CONSEQUENCES

A trust-native internet changes expectations.

Users stop proving who they are constantly. Systems stop asking for standing access. Consent becomes event-based rather than ambient.

Legitimacy becomes visible.

FROM PLATFORMS TO PROTOCOLS

Platforms centralize trust.

Protocols distribute it.

As authority becomes explicit and verifiable, protocols gain expressive power. Platforms shrink to interfaces rather than arbiters.

This is a reversal of recent history.

THE SHAPE OF WHAT FOLLOWS

A trust-native internet will not arrive fully formed.

It will emerge incrementally as systems adopt explicit authority, cryptographic intent, and bounded delegation. Each adoption reduces reliance on intermediaries.

Momentum builds quietly.

———

In the final chapter, we return to the individual, and the question of what dignity, agency, and responsibility mean in a world where authority is finally provable.

CHAPTER 28
DIGNITY IN A
PROVABLE WORLD

Technology is often judged by what it enables.

Less often by what it protects.

This book has argued that the most important shift underway is not faster systems, smarter agents, or stronger security. It is the emergence of proof as a first-class property of digital life.

That shift has human consequences.

DIGNITY AS A SYSTEM PROPERTY

Dignity is usually treated as a moral concern.

In practice, it is shaped by systems. When systems assume, infer, and observe by default, individuals are reduced to sources of data and targets of control.

When systems require explicit consent and provable authority, individuals retain agency.

Dignity becomes structural.

FROM BEING WATCHED TO BEING ASKED

Most digital interactions today are asymmetrical.

Systems watch continuously. Individuals are evaluated silently. Decisions are made elsewhere and revealed after the fact.

A decision-centric world reverses this posture. Systems ask at moments that matter. Individuals respond deliberately.

This shift restores balance.

RESPONSIBILITY MADE VISIBLE

Opacity erodes responsibility.

When authority is implicit, responsibility diffuses. Errors become systemic. Accountability becomes rhetorical.

Explicit authority makes responsibility legible. Decisions can be traced to commitments. Mistakes can be owned, corrected, and learned from.

This is not punitive. It is clarifying.

FREEDOM THROUGH CONSTRAINT

Unlimited systems are not liberating.

They are unpredictable.

By constraining when and how authority may be exercised, decision-centric systems create freedom elsewhere. Silence becomes the default. Intervention becomes meaningful.

Constraint protects autonomy.

TRUST WITHOUT SUBMISSION

Trust has often required submission.

Users surrender data. Organizations surrender control. Regulators surrender precision.

Proof changes the terms. Trust no longer requires visibility into everything. It requires evidence at the point of commitment.

This enables cooperation without domination.

A MORE HONEST RELATIONSHIP WITH MACHINES

As machines act more autonomously, the temptation is to anthropomorphize them.

To treat systems as decision-makers rather than instruments.

Explicit authority resists this drift. Machines execute. Humans decide. Proof records the boundary.

This honesty matters.

THE QUIET RESTORATION OF AGENCY

Most users will not notice when this transition occurs.

They will simply be interrupted less often. Asked fewer irrelevant questions. Monitored less continuously.

Agency returns quietly, as a reduction in friction and exposure.

BUILDING FOR THE PERSON, NOT THE PROFILE

Data-driven systems optimize for profiles.

Decision-driven systems respect persons.

This distinction determines whether technology amplifies individuality or compresses it into averages.

WHAT ENDURES

Protocols will change.

Cryptography will evolve.

What must endure is the principle that authority should be explicit, bounded, and provable, because people deserve to know when power is exercised over them, and to participate when it is.

THE FUTURE IS DELIBERATE

The future described in this book is not automated by default.

It is deliberate by design.

Systems move faster. Decisions become clearer. Privacy improves as a side effect of legitimacy.

This is not a return to the past. It is an upgrade in how we govern ourselves.

CLOSING THE LOOP

We began with a simple observation: modern systems can act, but they cannot decide.

By introducing explicit, cryptographically provable decisions, we restore a missing layer of digital life.

Not for machines.

For people.

EPILOGUE
LEGITIMACY AS INFRASTRUCTURE

This book has argued that trust, when treated as an assumption, becomes a liability.

The systems that now govern identity, access, allocation, and action operate at speeds and scales that make implicit legitimacy untenable. When authority is inferred rather than proven, power accumulates without accountability, and errors propagate faster than they can be corrected.

The response is not restraint alone, nor oversight after the fact, but architectural change.

WHAT HAS BEEN ESTABLISHED

If the arguments in this book hold, then several conclusions follow:

- Authority that cannot be proven will eventually be abused or contested.
- Auditability that relies on reconstruction will always arrive too late.

- Privacy erodes when systems compensate for uncertainty with data collection.
- Governance fails when legitimacy is implied rather than explicit.

Legitimacy must therefore become a first-class property of systems: not a legal interpretation, not a policy overlay, but an intrinsic, verifiable fact.

FROM DECISION TO PROOF

A system capable of proving who decided what, under which authority, at the moment of action does more than satisfy compliance or reduce risk.

It restores agency.

It allows consent to be intentional rather than assumed. It allows accountability to be precise rather than collective. And it enables trust to be established at the point of execution, not reconstructed after harm has occurred.

This is the shift from control to proof.

CONTINUITY

This book follows *The Hyperstore Manifesto*, which addressed the unification of data and execution.

Together, these works describe a Hypermodern condition: systems that scale faster than the structures meant to justify and govern them. Where the first book established a foundation for authoritative state, this one establishes a foundation for legitimate action.

The domains are distinct. The dependency is not.

WHAT FOLLOWS

Legitimacy is not an abstract ideal. It is an engineering problem.

The tools now exist to build systems where authority is explicit, consent is deliberate, and trust is earned through proof rather than habit. Whether those systems become the norm is no longer a question of feasibility, but of responsibility.

This book does not close that question.

It makes it unavoidable.

ABOUT THE AUTHOR

Dom Jocubeit is a systems architect and builder focused on the structural limits of modern digital infrastructure.

His work examines how authority, legitimacy, and coordination break down when systems scale faster than the mechanisms meant to govern them. Rather than treating trust, privacy, and governance as policy problems, he approaches them as architectural constraints that must be enforced mechanically.

Jocubeit is the creator of *PrivacyFirst.id*, a platform developed as a **falsifiable implementation** of the ideas explored in this book. His work builds on earlier efforts to unify data, execution, and authority into systems that can prove, not merely assert; correctness and legitimacy.

Across research, writing, and production systems, the author's focus is consistent: designing infrastructures where guarantees are structural, enforcement is intrinsic, and human discretion is replaced with executable constraint.

This book is part of *The Hypermodern Theorem*, a multi-volume exploration of data, authority, and coordination as foundational domains of trustworthy systems.

linkedin.com/in/jocubeit
x.com/jocubeit

GLOSSARY
HYPERPROOF TERMS

Authority
The formally recognized capacity to make a decision that has binding effect within a system. In this book, authority is not social, institutional, or assumed; it is explicit, bounded, and mechanically enforceable.

Assumed Trust
A design condition in which legitimacy is inferred from system participation (e.g., login, role, or process compliance) rather than proven at the moment of action.

Auditability (Structural)
The property of a system in which decisions are inherently inspectable and verifiable because proof is produced at execution time, not reconstructed after the fact.

Consent
An intentional, context-specific grant of authority that is explicit in scope and time. Consent is treated as an executable event, not an implied state or blanket agreement.

Control
Mechanisms that restrict or permit actions through policy, roles, or enforcement layers. Control differs from proof in that it limits behavior without necessarily establishing legitimacy.

Decision
A first-class object representing an intentional act taken under a specific authority and context. Decisions are distinct from side effects or automated outcomes.

Explicit Authority
Authority that is declared, scoped, and verifiable at the moment a decision is made, rather than inferred from identity, role, or historical access.

Falsifiable Implementation

A concrete system designed to test the claims of a manifesto or architectural thesis by exposing them to real-world constraints and potential failure.

Governance

The set of mechanisms by which authority, responsibility, and accountability are distributed and enforced within a system. In this book, governance is treated as an architectural problem, not a managerial or policy exercise.

Implicit Trust

Reliance on assumptions about legitimacy based on system boundaries, institutional reputation, or historical behavior rather than explicit proof.

Inference

The act of reconstructing legitimacy after an event has occurred using logs, policies, testimony, or interpretation. Inference is contrasted with proof.

Legitimacy

The condition in which an action can be shown to have been taken intentionally, by an authorized actor, within valid bounds, at the moment it occurred.

Policy

A declarative rule describing permitted or prohibited actions. Policies alone do not establish legitimacy unless coupled with explicit authority and proof.

Privacy

A systemic outcome that arises when authority, consent, and decision-making are explicit and bounded. Privacy is not treated as a feature or add-on.

Proof

Cryptographic or structural evidence produced at execution time that establishes who decided what, under which authority, and in what context.

Surveillance

The compensatory over-collection and monitoring of data that occurs when systems cannot prove legitimacy and instead rely on inference.

Trust
A social workaround for systems that cannot enforce or prove their own rules. In this book, trust is treated as a liability at scale.

Verifiable Decision-Making
A system property in which every consequential action is accompanied by proof of authority, intent, and scope at the time of execution.

www.ingramcontent.com/pod-product-compliance
Lightning Source LLC
Chambersburg PA
CBHW070522200326
41519CB00013B/2891